Aleksander Annov
Anton Annov

Average chords and diameters of the ellipse

AF153252

Aleksander Annov
Anton Annov

Average chords and diameters of the ellipse

Computer simulation and results of computing experiments

LAP LAMBERT Academic Publishing

Impressum / Imprint
Bibliografische Information der Deutschen Nationalbibliothek: Die Deutsche Nationalbibliothek verzeichnet diese Publikation in der Deutschen Nationalbibliografie; detaillierte bibliografische Daten sind im Internet über http://dnb.d-nb.de abrufbar.
Alle in diesem Buch genannten Marken und Produktnamen unterliegen warenzeichen-, marken- oder patentrechtlichem Schutz bzw. sind Warenzeichen oder eingetragene Warenzeichen der jeweiligen Inhaber. Die Wiedergabe von Marken, Produktnamen, Gebrauchsnamen, Handelsnamen, Warenbezeichnungen u.s.w. in diesem Werk berechtigt auch ohne besondere Kennzeichnung nicht zu der Annahme, dass solche Namen im Sinne der Warenzeichen- und Markenschutzgesetzgebung als frei zu betrachten wären und daher von jedermann benutzt werden dürften.

Bibliographic information published by the Deutsche Nationalbibliothek: The Deutsche Nationalbibliothek lists this publication in the Deutsche Nationalbibliografie; detailed bibliographic data are available in the Internet at http://dnb.d-nb.de.
Any brand names and product names mentioned in this book are subject to trademark, brand or patent protection and are trademarks or registered trademarks of their respective holders. The use of brand names, product names, common names, trade names, product descriptions etc. even without a particular marking in this works is in no way to be construed to mean that such names may be regarded as unrestricted in respect of trademark and brand protection legislation and could thus be used by anyone.

Coverbild / Cover image: www.ingimage.com

Verlag / Publisher:
LAP LAMBERT Academic Publishing
ist ein Imprint der / is a trademark of
OmniScriptum GmbH & Co. KG
Heinrich-Böcking-Str. 6-8, 66121 Saarbrücken, Deutschland / Germany
Email: info@lap-publishing.com

Herstellung: siehe letzte Seite /
Printed at: see last page
ISBN: 978-3-659-51760-0

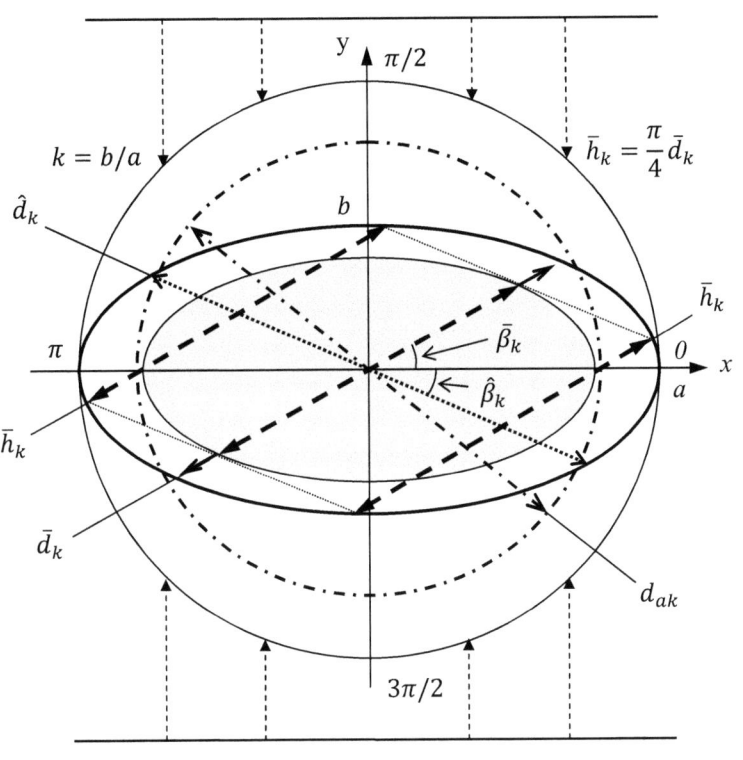

CONTENTS

4

The accepted symbols

k — compression ratio of the ellipse, shares of units;

$d_{\beta/x} = d_\beta/d_x$ - current relative diameter of the ellipse, at which the angle of rotation (β) or the slope $\varphi_\beta = \tan\beta$;

$d_{(\beta\pm\frac{\pi}{2})/x}$ — relative diameter, perpendicular to diameter ($d_{\beta/x}$) and which angle of rotation from the major bigger axis of the ellipse is equal $(\beta \pm \pi/2)$;

$d_{ak/x}$ — equivalent relative diameter of the ellipse, which compression ratio (k);

$Q\{h_{\beta i/x}\}, d_{\beta/x} \in Q$ — a set of parallel relative chords ($h_{\beta i/x}$), including diameter ($d_{\beta/x}$), which current angle of rotation relative bigger major diameter (d_x) is equal (β), further designated as $Q\{h_{\beta i/x}\}$;

$\tilde{Q}\{h_{k\beta i/x}\}$ — stepwise rotating variable set of parallel relative chords of the ellipse, including diameter ($d_{\beta/x}$), which variable current angle of rotation relative bigger major diameter (d_x) is equal (β), at a rotation step ($\Delta\beta$) and compression ratio (k);

$h_{\beta i/x}$ — current (i) a relative chord of the ellipse in a set of $Q\{h_{\beta i/x}\}$;

$\bar{h}_{\beta/x}$ — the average relative chord of a set of $Q\{h_{\beta i/x}\}$;

$\bar{d}_{k/x}$ — the average relative diameter of the ellipse, which compression ratio (k);

$\pm\bar{\varphi}_k, \pm\bar{\beta}_k$ — angular coefficients (slopes) and angles of rotation of two, symmetrically located, averages ($\bar{d}_{k/x}$) diameters of the ellipse, in radians;

$\hat{d}_{k/x}$ — the relative diameter, conjugate to set of parallel chords, including the average relative diameter ($\bar{d}_{k/x}$) of the ellipse;

$\Delta h_x = \Delta h/d_x$ — a relative step (distance) between parallel chords in a set of $Q\{h_{\beta i/x}\}$, shares of units;

$\overrightarrow{\Delta h}_x$ — the specified limit relative step between parallel chords, shares of units;

$\sum_{\Delta h_x} H_{\beta i/x}$ — the total length of the set $Q\{h_{\beta i/x}\}$ parallel chords, of a definite step (Δh) between parallel chords;

$\sum_{\Delta \vec{h}_x} H_{\beta i/x}$ − limit value of the sum of the set $Q\{h_{\beta i/x}\}$ parallel chords of one ellipse, at the specified limit step $(\Delta \vec{h}_x)$ between chords;

$\sum_{\Delta \vec{h}_x} H_{k/x}$ − limit value of the sum of all massif of chords of the ellipse, which compression ratio (k);

$n_{\beta,\Delta \vec{h}_x}$ − limit number of chords in a set of $Q\{h_{\beta i/x}\}$, at the specified limit step $(\Delta \vec{h}_x)$ between chords;

$n_{k,\Delta \vec{h}_x}$ − limit number of chords in the general massif of all chords of the ellipse, which compression ratio (k), at the specified limit step $(\Delta \vec{h}_x)$ between chords;

$\tilde{n}_\beta$ − variable number of parallel chords in a rotating set of $\tilde{Q}\{h_{\beta i/x}\}$, depending on the current rotation angle (β) and at the specified limit of relative step $(\Delta \vec{h}_x)$ between chords;

$S_{k/x^2} = S_k/d_x^2$ − the relative area of the ellipse, which compression ratio (k).

1. Introduction

In some cases there is a need to recognize the shape and size of figures, the shape close to a rectangle, circle or oval. To simplify the task of recognition of such figures is considered closest to him in the form of a figure - an ellipse. As the main characteristic parameters considered diameters and chords of the ellipse and the characteristic relationships between them.

For calculation of diameters and chords of ellipses and their ratios, the method of computer simulation is applied. With the support of Excel 2007, is considered a model stepwise rotating variable set of parallel chords on entire range of forms of ellipses $k = [0; 1]$. In view the symmetry of the ellipse, considered a range of rotation sets the interval $\beta = [0; \pi]$.

All key parameters, for the purpose of their universalization for any sizes of ellipses, are considered in relative units (RLU), by division of their sizes into bigger major (d_x) diameter of an ellipse.

All average values of chords and diameters are arithmetic averages.

1.1. The compression ratio of the ellipse

It is well-known that at uniform compression (*Fig. 1*), the circle will be transformed to an ellipse.

Extent of compression [1⁵⁶] is characterized by compression ratio $(k = b/a)$, and expression $(1 - k) = (a - b)/a$ is called as ellipse compression, where (a) и (b) − semi-axis of the ellipse. The axis of compression (x) is called as a big axis of an ellipse, and perpendicular as it an axis (y), passing through the center of an ellipse, is called as a small axis of an ellipse.

2. The diameters of the ellipse

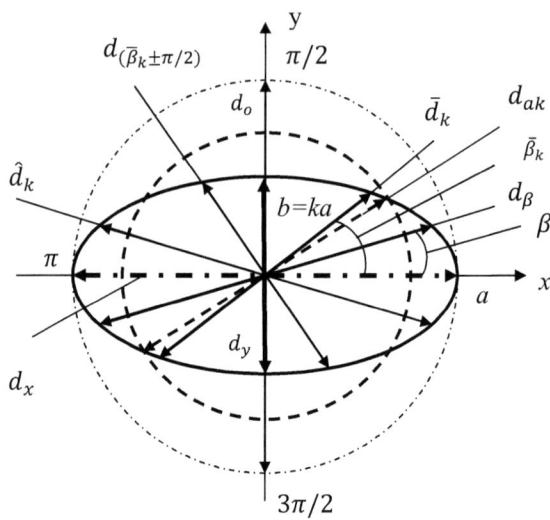

Fig. 1. The diameters of the ellipse, resulting in a uniform compression in the range of $(1/k)$ times, where:

d_o — diameter of the original circle;

d_x и d_y — major diameter of the ellipse, and $d_x = d_o = 2a$, $d_y = 2b$;

d_β — one of a plurality of possible diameters of the ellipse, angle of rotation relative to the major axis of the ellipse is (β), or the slope is $\varphi_\beta = \tan\beta$;

d_{ak} — equivalent diameter of the ellipse with a compression ratio of (k), or diameter of a circle whose area is equal to the area of the ellipse;

$\bar{d}_k$ — the average diameter of the ellipse, with a compression ratio (k) and the rotation angle $(\bar{\beta}_k)$;

$d_{(\bar{\beta}_k \pm \pi/2)}$ — the diameter perpendicular to the average diameter $(\bar{d}_k)$ of the ellipse;

$\hat{d}_k$ — diameter, conjugate (passing through the middle of the chords) to set of parallel chords, including the average diameter $(\bar{d}_k)$ of the ellipse.

2.1. The average diameters of the ellipse

It is known [1 503] line equation (r_β), extending from the center of the ellipse to the point (M) of intersection of the ellipse

$$r_\beta = \frac{ab}{\sqrt{b^2\cos^2\beta + a^2\sin^2\beta}};$$ (1)

where: β — angle between the line (r_β) and the x - axis.

In the transition to the relative diameters of the ellipse and the use of known trigonometric identities $(1 + \tan^2\beta = 1/\cos^2\beta; \quad 1 + \cot^2\beta = 1/\sin^2\beta)$, equation diameter of the ellipse is given by

$$d_{\beta/x} = \frac{k}{\sqrt{k^2\cos^2\beta + \sin^2\beta}} = k\sqrt{\frac{1+\tan^2\beta}{k^2+\tan^2\beta}} = k\sqrt{\frac{1+\varphi_\beta^2}{k^2+\varphi_\beta^2}}.$$ (2)

The result of calculation (2) using the *Excel* [2], the steps of rotating relative diameter $(d_{\beta/x})$ of the ellipse, with the step of rotation $(\Delta\beta = 0{,}01\pi)$ and interval of rotation angle at $\beta = [0; \pi]$, obtained by family of curves $d_{\beta/x} = f(k)$ changes relative diameter $d_{\beta/x}$, for different values of (k). The interval $\beta = [0; \pi]$ the angle of rotation diameter available *Fig. 1* is one diameter coinciding with the direction of major diameters of the ellipse (d_x, d_y), and a pair of other diameters. The slope and the angles of rotation of pairs of equal diameters have different signs $(\pm\varphi_\beta, \pm\beta)$. The family consists of the 40 curves resulting from the decomposition step $\Delta k = 0{,}025$ total interval $k = [0; 1]$. On *Fig. 2* shows a selection from this family, where the upper horizontal line at the level of $(d_{\beta/x} = 1)$ corresponds to a circle $(k = 1)$. By increasing the compression range $(k \to 0)$, is increased deflection curves of relative diameter and, for $(k = 0{,}025)$, the relative diameter of the ellipse, within the boundaries of the interval the angle of rotation $\beta = [0; \pi]$, is $d_{\beta/x} \cong 0$.

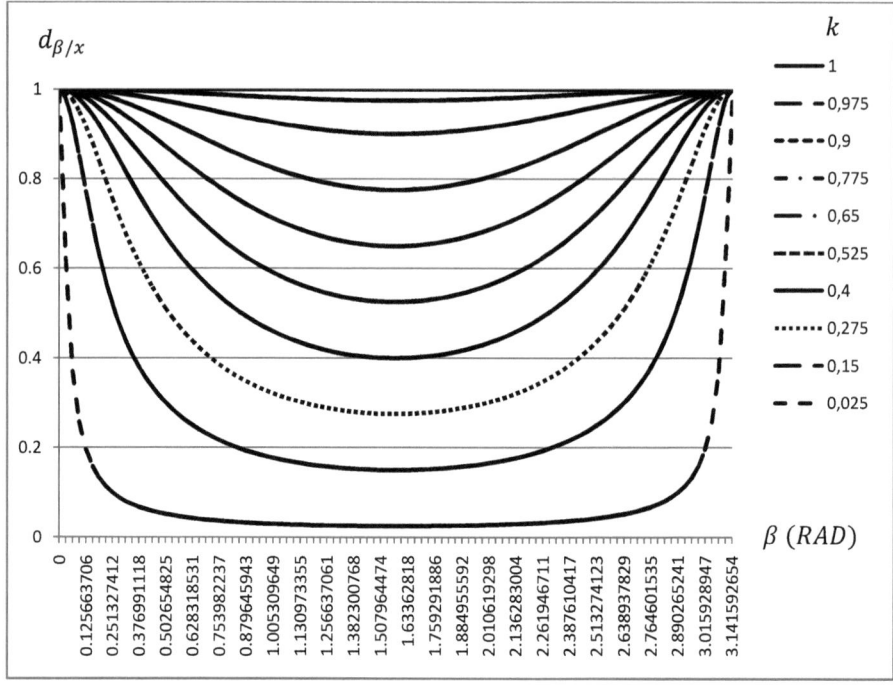

Fig.2. Sample from a family of curves changes relative diameter $d_{\beta/x}$ of the ellipse on the interval the angle rotation $\beta = [0; \pi]$, for different values of $k = [0; 1]$.

For each curve of the family corresponding to a specific compression ratio, defined value [2] relative average diameter ($\bar{d}_{k/x}$). Collectively, these values after the interpolation function of the relative average diameter $\bar{d}_{k/x} = f(k)$ of the ellipse, presented in summary *Table 2* and *Fig. 7*. Ibid presented only positive values of the angular coefficients $\pm\bar{\varphi}_k = f(k)$ and rotation angles $\pm\bar{\beta}_k = f(k)$ in radians, whose calculation is made taking into account (2) by the following formulas

$$\bar{d}_{k/x} = k\sqrt{\frac{1+tan^2(\pm\bar{\beta}_k)}{k^2+tan^2(\pm\bar{\beta}_k)}} = k\sqrt{\frac{1+(\pm\bar{\varphi}_k)^2}{k^2+(\pm\bar{\varphi}_k)^2}}; \quad (3)$$

$$\bar{\varphi}_k = \pm k\sqrt{\frac{d_x^2-\bar{d}_k^2}{\bar{d}_k^2-k^2 d_x^2}} = \pm k\sqrt{\frac{1-(\bar{d}_{k/x})^2}{(\bar{d}_{k/x})^2-k^2}}; \quad (4)$$

$$\bar{\beta}_k = tan^{-1}(\pm\bar{\varphi}_k) = \pm tan^{-1}(\bar{\varphi}_k). \quad (5)$$

The sign ($\pm$) indicates the presence of two average diameters of an ellipse.

In the future, since the symmetry of the ellipse, average relative diameters of the ellipses are considered only with positive angles.

The angles of rotation ($\bar{\beta}_k^o$) of average diameters of the ellipse in degrees, shown in *Table 3* and *Fig. 8*.

2.2. The equivalent diameter of the ellipse

Well known formula for calculating the diameter (d_{ak}) of circle equivalent, obtained by comparison of equal size area, the ellipse and the circle (*Fig. 1*)

$$d_{ak} = \sqrt{4ab} = \sqrt{d_x d_y} = d_x\sqrt{k}, \quad \text{or} \quad d_{ak/x} = \sqrt{k}. \tag{6}$$

The function of the relative equivalent diameter of the ellipse $d_{ak/x} = f(k)$, calculated [2] of formula (6) to the interval $k = [0; 1]$, is presented in *Table 2* and in *Fig. 7*. There are also the results of calculations of slopes $\varphi_{ak} = f(k)$ and angles of rotation $\beta_{ak} = f(k)$ in radians.

Angles of rotation (β_{ak}^o) of the equivalent diameters in degrees, shown in *Table 3* and *Fig. 8*.

The ratio of the average ($\bar{d}_k$) diameter (3) of the ellipse to the equivalent (d_{ak}) diameter (6) has the form

$$\frac{\bar{d}_k}{d_{ak}} = \frac{\bar{d}_{k/x}}{d_{ak/x}} = \sqrt{k\left(\frac{1+tan^2\bar{\beta}_k}{k^2+tan^2\bar{\beta}_k}\right)} = \sqrt{k\left(\frac{1+\bar{\varphi}_k^2}{k^2+\bar{\varphi}_k^2}\right)}. \tag{7}$$

From *Table 2*, *Fig. 7* and formula (7) that average diameter not equal an equivalent diameter of the ellipse, it is less than it. Their slopes and angles of rotation are not equal.

2.3. The conjugate diameter of the ellipse

In parallel, the average relative diameter ($\bar{d}_{k/x}$) of the ellipse can spend an endless set $Q\{h_{k\beta i/x}\}$ chords. Is well known [1^{76}] the relationship between the angular coefficients of average diameter $(\bar{d}_{k/x})$ the set of parallel chords of $Q\{h_{k\beta i/x}\}$ and

the diameter $(\hat{d}_{k/x})$, the conjugate of this set

$$\hat{\varphi}_k = -k^2/\bar{\varphi}_k, \tag{8}$$

where: $\bar{\varphi}_k$ — the slope of the average diameter $(\bar{d}_{k/x})$ of the ellipse, with compression ratio (k);

$\hat{\varphi}_k$ — the slope of the conjugate diameter $(\hat{d}_{k/x})$.

The relative value of the conjugate diameter $(\hat{d}_{k/x})$, according to (2) and (8), defined by the expression

$$\hat{d}_{k/x} = \frac{\hat{d}_k}{d_x} = k\sqrt{\frac{1+\hat{\varphi}_k{}^2}{k^2+\hat{\varphi}_k{}^2}} = \sqrt{\frac{k^4+\bar{\varphi}_k{}^2}{k^2+\bar{\varphi}_k{}^2}}. \tag{9}$$

The calculation results [2] of $\hat{d}_{k/x} = f(k)$ are presented in *Table 2* and in *Fig. 7*.

Angle $(\hat{\beta}_k)$ rotating of the conjugate diameter $(\hat{d}_{k/x})$, about the x — axis, in radians

$$\hat{\beta}_k = \tan^{-1}\hat{\varphi}_k = -\tan^{-1}(k^2/\bar{\varphi}_k). \tag{10}$$

The calculation results [2] functions $\hat{\varphi}_k = f(k)$, $\hat{\beta}_k = f(k)$ и $\hat{\beta}_k^o = f(k)$ are presented in *Table 3* and *Fig. 8*.

Example. There is an ellipse with parameters:

$d_x = 2a = 8mm, d_y = 2b = 4mm.$

From *Tables 3* and *4*, for $k = b/a = 0{,}5$:

$\bar{d}_{k/x} = 0{,}68644;$	$\hat{d}_{k/x} = 0{,}8825;$	
$\bar{d}_{k/x} \cdot d_x = 5{,}49mm;$	$\bar{\varphi}_k = 0{,}77307737;$	$\bar{\beta}_k^o = 37{,}707°;$
$\hat{d}_k = \hat{d}_{k/x} \cdot d_x = 7{,}06mm;$	$\hat{\varphi}_k = -0{,}32338289;$	$\hat{\beta}_k^o = -17{,}92°.$

3. The chords of the ellipse

3.1. Variable rotating set of parallel chords of an ellipse

Any parallel relative to the diameter $(d_{\beta i/x})$ of the ellipse can spend uncountable (endless) set $R\{h_{\beta i/x}\}$ of chords $(h_{\beta i/x})$, which is the range of values in the interval $[0; d_{\beta/x}]$, and the value $(h_{\beta i/x})$ is a continuous random variable. Continuous random variable $(h_{\beta i/x})$ is transformed into a discrete, when converting an uncountable set $R\{h_{\beta i/x}\}$ of chords in a countable (finite) set $Q\{h_{\beta i/x}\}$, by setting fixed step (Δh_x) between parallel chords (*Fig. 3*). Stepping rotation of such a set , with a step rotation $(\Delta\beta)$, varying the number of countable $(\tilde{n}_\beta)$ parallel chords , with rotation angles $\beta = (0 ; \pi / 2 ; \pi; 2\pi / 3 , ...)$, is given by $(\tilde{n}_\beta = d_{(\beta\pm\frac{\pi}{2})/x}/\Delta h_x)$.

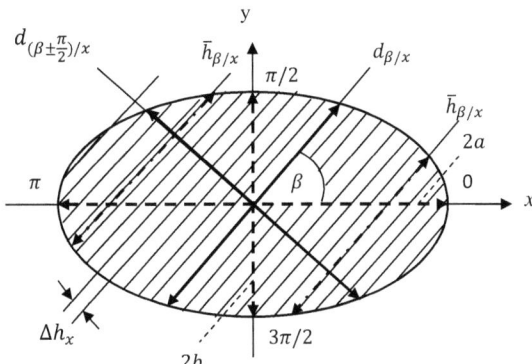

Fig.3. Finite set $Q\{h_{\beta i/x}\}$ parallel relative chords of the ellipse with the relative step (Δh_x) between chords, one of which is the diameter $(d_{\beta/x})$, and the other two $(\bar{h}_{\beta/x})$ average chord sets. Leftmost chord intersects diameter $(d_{(\beta\pm\frac{\pi}{2})/x})$ outside the ellipse.

As follows from *Fig. 3*, at other angles calculation of $(\tilde{n}_\beta)$ this formula is not exact due to the fact that part of the current set of chords $Q\{h_{\beta i/x}\}$ crosses the line diameter $d_{(\beta\pm\frac{\pi}{2})/x}$ outside the ellipse and not considered by the formula. This problem is avoided by the computer simulation, stepwise rotating variable set of the parallel chords.

3.2. Quasi-endless set of parallel chords of an ellipse

It is known [1483] that the curvilinear figure decomposition parallel chords other set of curvilinear trapezoids width (Δh), the sum of the areas of all trapezoids equal to the area of the figure. Well-known theorem [1441], explaining the essence of the definite integral, "Partitioning figures on indefinitely increasing the number of intervals, the sum product of the width intervals for their height tends to a limit, which is constant for any method of partial formation interval."

With respect to the ellipse, which has a finite set of parallel chords $Q\{h_{\beta i/x}\}$ with a fixed rotation angle (β), the relative area of the ellipse (S_{k/x^2}) is equal to

$$S_{k/x^2} = \Delta h_x \sum_{\Delta h_x} H_{\beta i/x} \pm \Delta S_{k/x} \tag{11}$$

For endless sets of parallel chords, at $\Delta h_x \to 0$, error ($\Delta S_{k/x} \to 0$), and the sum of the areas of all curvilinear trapezoids formed by set of $Q\{h_{\beta i/x}\}$, is committed to the area (S_{k/x^2}) of the ellipse

$$\{\lim_{\Delta h_x \to 0} \Delta h_x \sum_{\Delta h_x} H_{\beta i/x}\} = S_{k/x^2}; \tag{12}$$

According to the above mentioned theorem [1441], at $\Delta h_x \to 0$, the sum of ($\sum_{\Delta h_x} H_{\beta i/x}$) chords stepwise rotating the set $\tilde{Q}\{h_{k\beta i/x}\}$, is constant in all directions, for all values of (β), which is confirmed by the results of [2] computer simulation.

For endless sets, taking into account (12) and known [1485] formulas ($S_k = \pi ab = \pi k a^2 = \pi k \frac{d_x^2}{4}$):

$$S_{k/x^2} = \{\lim_{\Delta h_x \to 0} \Delta h_x (\sum_{\Delta h_x} H_{\beta i/x})\} = \frac{\pi}{4} k; \tag{13}$$

From the (13) that if we take the step value between the chords close but not equal to zero ($\Delta \vec{h}_x \cong 0$), for example ($\Delta \vec{h}_x = 0,0005$), it is permissible error is small ($\Delta S_{k/x} \to 0$), can be neglected. And thus, the conditional pass from the finite to the quasi-endless set of parallel chords.

3.2.1. The area and the total length of chords in the quasi-endless set

For quasi-endless sets, when $\Delta h_x \to (\Delta \vec{h}_x \cong 0)$, expression (13) takes the form

$$\lim_{(\Delta h_x \to \Delta \vec{h}_x)} \Delta h_x (\sum_{\Delta h_x} H_{\beta i/x}) = \Delta \vec{h}_x \sum_{\Delta \vec{h}_x} H_{\beta i/x} = \frac{\pi}{4} k. \tag{14}$$

From which we obtain a formula for the calculation of the limiting values of the relative total length of quasi-endless set of parallel relative chords of the ellipse, given the limiting relative step between chords.

Whence
$$\sum_{\Delta \vec{h}_x} H_{\beta i/x} = \frac{\pi}{4} \frac{k}{\Delta \vec{h}_x}. \tag{15}$$

The limiting value relative total length quasi-endless set of parallel relative chords of the ellipse, given the limiting relative step between chords, depends only on specified limiting step between chords and the compression ratio of the ellipse. It is independent on the angle of rotation of the set and continuously in all directions, as evidenced by the results of computer simulation [2] and is consistent with the theorem [1^{441}].

In *Table 1* and *Figure 4,5* are linear functions $\sum H_{\beta i/x (\Delta \vec{h}_x = 0,0005)} = f(k)$ and $S_{k/x^2 (\Delta \vec{h}_x = 0,0005)} = f(k)$, obtained by computer simulation of quasi-endless sets of parallel chords. According to (13), the area of the ellipse for such sets coincides with the actual area of the ellipse.

Table 1

k	$\sum H_{\beta i/x (\Delta \vec{h}_x = 0,0005)}$	$S_{k/x^2 (\Delta \vec{h}_x = 0,0005)}$
0	0	0
0,1	157,0796327	0,078539816
0,2	314,1592654	0,157079633
0,3	471,2388982	0,235619449
0,4	628,3185307	0,314159265
0,5	785,3981634	0,392699082
0,6	942,4777961	0,471238898
0,7	1099,5574290	0,549778714
0,8	1256,6370611	0,628318531
0,9	1413,7166942	0,706858347
1,0	1570,7963272	0,785398163

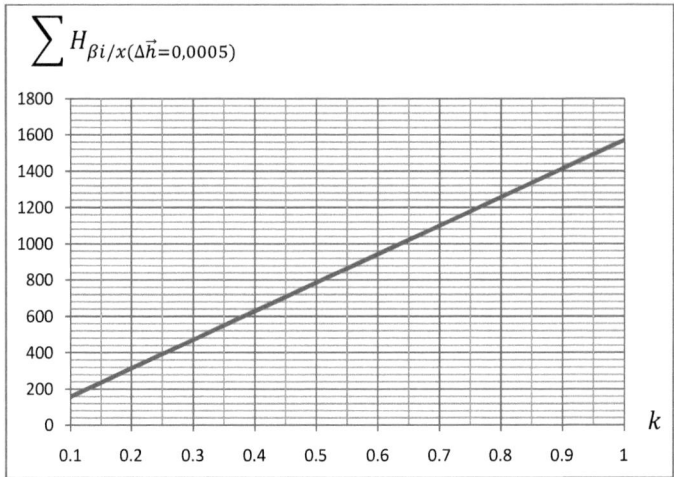

Fig.4. Changing the limit of relative total length of quasi-endless set of parallel relative chords of the ellipse, at specified the limiting relative step between the chords ($\Delta \vec{h}_x = 0{,}0005$), depending on changes in the compression ratio (k) of the ellipse. The total length of the set of parallel chords does not depend on the angle of rotation of the set; it is constant in all directions.

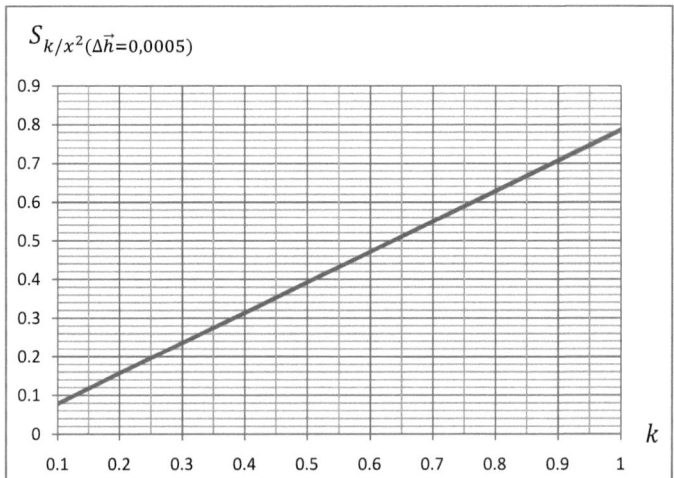

Fig.5. Changing the calculation of the relative area of the ellipse, specified the relative limiting step ($\Delta \vec{h}_x = 0{,}0005$) between chords, depending on changes in the compression ratio (k) of the ellipse.

3.2.2. *Example of calculating the area and the total length of the chords*

Consider an ellipse with a compression ratio $k = 0,5$ and size $d_x = 10mm$.

In the ellipse has a quasi-endless set of parallel relative chords with the specified relative step between the chords ($\Delta \vec{h}_x = 0,0005$), thus achieving ($\Delta S_{k/x} \cong 0$).

According to (15), the limit value of the relative overall length of the set of parallel relative chords of the ellipse to each direction:

$$\Sigma_{(k=0,5;\ \Delta \vec{h}_x=0,0005)}\ H_{\beta i/x} = \frac{\pi k}{4\Delta \vec{h}_x} = 785,398, \text{RLU (relative units)}.$$

The total length of the set of chords: $\qquad \Sigma H_{\beta i} = d_x (\Sigma H_{\beta i/x}) = 7854\ mm.$

Relative area of the ellipse (13): $\qquad S_{k/x^2} = \frac{\pi}{4}k = 0,392699\ \text{RLU}.$

The real area of an ellipse: $\qquad S_k = \frac{\pi}{4}kd_x{}^2 = 39,27\ mm^2.$

This result is confirmed by the results of [2] computer simulation.

Thus, the results of the considered model of the quasi-endless set of parallel chords, can be roughly, at an insignificant error, relate to infinite sets of parallel chords of an ellipse.

3.3. *Computer simulation of sets of parallel chords of an ellipse*

3.3.1. *Average chord sets of parallel chords of an ellipse*

In any set of $Q\{h_{\beta i/x}\}, d_{\beta/x} \in Q$ parallel chords of an ellipse (*Fig. 3*) there are two, equal in value, the average chord ($\bar{h}_{\beta/x}$). Their relative value is determined by the expression

$$\bar{h}_{\beta/x} = \Sigma_{\Delta \vec{h}_x} H_{\beta i/x} / \tilde{n}_\beta. \tag{16}$$

Whence, by (15), a variable number of parallel chords in a rotating set of $\tilde{Q}\{h_{\beta i/x}\}$, depending on the rotation angle (β) the set

$$\tilde{n}_\beta = \sum \Delta\vec{h}_x \, H_{\beta i/x} / \overline{h}_{\beta/x} = \pi k / 4\Delta\vec{h}_x \, \overline{h}_{\beta/x}; \qquad (17)$$

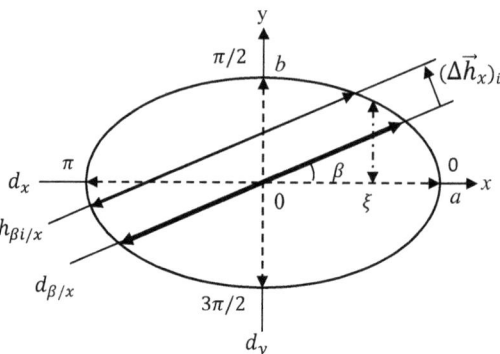

Fig.6. Parallel shift once (i) the relative chord ($h_{\beta i/x}$) from the co-rotational relative diameter ($d_{\beta/x}$) for a specified value of the limiting step ($\Delta\vec{h}_x$) relative displacement.

3.3.2. Simulation algorithm

For precise determination of the average relative chord ($\overline{h}_{\beta/x}$) in each set of parallel chords of $Q\{h_{\beta i/x}\}$ ellipses any shape and size, the method of computer simulation, stepwise rotating variable set $\tilde{Q}\{h_{k\beta i/x}\}$, using the *Excel*, allowing to define the standard statistical parameters . Simulation each set with a specific rotation angle (β) is (*Fig. 6*) in parallel stepwise increasing shift next (i) the relative chord ($h_{\beta i/x}$) from the co-rotational relative diameter ($d_{\beta/x}$) for a specified value of the limiting step ($\Delta\vec{h}_x$) relative displacement.

Program *Excel*, according to a specially prepared formula (18), calculates the relative length, next parallel and steps - displaceable segment between two points of intersection of the chord with the ellipse

$$h_{\beta i/x} = \frac{h_{\beta i}}{d_x} = \frac{d_{\beta/x}}{k}\sqrt{k^2 - d_{\beta/x}^{\,2}(i\overrightarrow{\Delta h_a})^2} = \frac{d_{\beta/x}}{k}\sqrt{k^2 - d_{\beta/x}^{\,2}(2i\overrightarrow{\Delta h_x})^2}; \qquad (18)$$

where: a — semi-major axis of the ellipse, $d_x = 2\,a$;

i — sequence number of the next displaceable parallel to the chord.

According to (18), the value of the relative displacement $(i\overrightarrow{\Delta h_a} = 2i\overrightarrow{\Delta h_x})$ increased stepwise from zero to the appearance under the radical sign of a negative number [2]. A negative number is displayed when the next calculated chord is beyond the ellipse. Stepwise transition from one set of $Q\{h_{\beta i/x}\}$ for the subsequent, performed by rotating step at a rotation angle $(\Delta\beta)$. Interval stepwise rotations, view of the symmetry of ellipses, adopted $\beta = [0;\ \pi]$. In a way step through rotating variable set $\tilde{Q}\{h_{k\beta i/x}\}$ of parallel chords, are variable parameters $(\tilde{n}_\beta)$ and $(\overline{h}_{\beta i/x})$.

For comparison purposes, separately considered quasi-endless (conditionally endless) sets, under $(\overrightarrow{\Delta h_a} = 0{,}001;\ \Delta\beta = \pi/36)$, and finite sets of parallel chords, at $(\overrightarrow{\Delta h_a} = 0{,}01;\ \Delta\beta = \pi/18)$. The accuracy of calculations quasi-endless sets at step $(\overrightarrow{\Delta h_a} = 0{,}001)$, is acceptable for the generalization of the results, and it is slightly different from the precision in the step $(\overrightarrow{\Delta h_a} = 0{,}0005)$, in which significantly increases the amount of computation.

Simulation is performed in a certain sequence according to the defined algorithm. For each operation, in accordance with the formula used, prepared the corresponding programmable function (f_x).

Simulation algorithm, step through rotating set of parallel chords and average chords following definition:

- Setting the current value of the compression ratio (k) of the ellipse of a series of values $(0{,}1;\ 0{,}2;\ 0{,}3;\ 0{,}4;\ 0{,}5;\ 0{,}6;\ 0{,}7;\ 0{,}8;\ 0{,}9;\ 1)$; $f_x = \$A\$3$.

- Setting stepwise rotation angle (β) the set of parallel chords of an ellipse. At step rotation angle $(\Delta\beta = \pi/36)$, treated $(n = 36)$ positions the angles of rotation with addresses $(C1: AL1)$ and 36, the respective positions, rotation angles (β) step through rotating set of parallel chords with addresses $(C2: AL2)$, programmable cyclic function $f_x = C\$2 * PI()/36$. For finite sets of steps ($\Delta\beta = \pi/18$), respectively: $(C1:T1)$, $(C2:T2)$, $f_x = C\$2 * PI()/18$.

- Setting the limit step ($\Delta\vec{h}_a = 2\Delta\vec{h}_x = 0,001$) relative displacement next chord. Given that the limiting number of chords, calculated on one side of the co-rotational diameter $\vec{\iota} = 1/\Delta\vec{h}_a = 1000$, the maximum possible number of chords at each incremental set 2000 and, considering the diameter, of the total number 2001. Cell ($B4$) is set to 1, programmable cyclic function for cell $B5$: $f_x = (\$B4 - 0,001)$, which fills the cell cycle ($B4:B2004$). Cell ($B1005:B2004$), following the cell with the value zero ($B1004$), are negative. For finite sets: the maximum possible number of chords 201, a cell with a value of zero ($B104$), programmable function $f_x = (\$B4 - 0,01)$.

- Rows (1004) or (104), corresponding to a zero displacement from the diameter, filled with the values step through rotating diameter (2). Cyclic function of the cell ($D1004$) or, for finite sets, cell ($D104$):

f_x =$A\$3/(ROOT((($A\$3*$A\$3)*(COS(C\$2)*COS(C\$2)))+(SIN(C\$2)*SIN(C\$2))))).

- To calculate all the chords (18), step though rotating set of parallel chords, except diameters, cyclic function of the cell ($C4$)

f_x =(C\$1004/$A\$3)*(ROOT(($A\$3*$A\$3)-(C\$1004*C\$1004)*($B4*$B4))).

For finite sets the cell number (1004) is changed to (104).

- Table columns is defined of relative total length ($\sum_{\Delta\vec{h}_x} H_{\beta i/x}$) and the total number ($n_{\beta,\Delta\vec{h}_x}$) parallel relative chords in the current set of $Q\{h_{\beta i/x}\}$, including the diameter, and the arithmetic average chord ($\bar{h}_{\beta/x}$) set. The results are entered into the appropriate cells:

for quasi-endless sets ($C2007:AL2007, C2011:AL2011, C2013:AL2013$),

for finite sets ($C207:T207, C211:T211, C213:T213$).

- On row ($C1004:AL1004$) table or of finite sets on row ($C104:T104$) are determined by the average diameter ($\bar{d}_{k/x}$) of the ellipse.

- On row $(C2013: AL2013)$ or $(C213: T213)$ is determined by the arithmetic average of average chord of an ellipse $(\bar{\bar{h}}_{k\beta/x})$.

- By dividing the sum of all the chords $(\sum_{\Delta \bar{h}_x} H_{k/x})$, strings (C2007: AL2007) or (C207: T207), on their quantity $(n_{k,\Delta \bar{h}_x})$, strings (C2011: AL2011) or (C211:T211), is determined by the arithmetic average $(\widehat{\bar{h}}_{k/x})$ of statistical set of chords of the ellipse.

3.3.3. Simulation results

Calculation tables and calculations of finite sets, in the form of sample of separate tables are presented in *Appendix*.

Calculated table quasi-endless sets, due to their large size, are only present at the website [2]. The calculation results are quasi-endless sets, after interpolation, are shown in *Tables 2, 3* and *Fig. 7, 8*.

In Figure 7 shows graphs of the following functions:

$\bar{d}_{k/x} = f(k)$ − average relative diameter of the ellipse;

$d_{ak/x} = f(k)$ − equivalent relative diameter of the ellipse;

$\hat{d}_{k/x} = f(k)$ − relative diameter, conjugate to set of parallel chords, including the average relative diameter of the ellipse;

$\bar{\varphi}_k = f(k)$, $\bar{\beta}_k = f(k)$ − the slope and the angle of rotation of one of the two average diameters of the ellipse, in radians;

$\varphi_{ak} = f(k)$, $\beta_{ak} = f(k)$ − the slope and the angle of rotation of the equivalent diameter, in radians;

$\bar{\bar{h}}_{k\beta/x} = f(k)$ − arithmetic average of average chord, rotating variable set of parallel relative chords of an ellipse;

$\widehat{\bar{h}}_{k/x} = f(k)$ − arithmetic average of the statistical set of relative chords of an ellipse.

Table 2. Change the characteristic diameter (d), their angular coefficient (φ) and angles of rotation (β) for different compression ratios (k) of the ellipse. (Summary Table).

k	$d_{ak/x}$	φ_{ak}	β_{ak}	$\bar{d}_{k/x}$	$\bar{\varphi}_k$	$\bar{\beta}_k$	$\hat{d}_{k\beta/x}$	$\bar{\bar{h}}_{k\beta/x}$	$\widehat{h}_{k/x}$
	0	0	0	0	0	0	1	0	0
0,025	0,15811	0,1581	0,1568	0,08607	0,3024	0,2937	0,99661	0,06761	0,03081
0,050	0,22361	0,2236	0,2211	0,14068	0,3764	0,3611	0,99132	0,11049	0,06139
0,075	0,27386	0,2739	0,2673	0,19007	0,4216	0,3991	0,98463	0,14928	0,09214
0,100	0,31623	0,3162	0,3063	0,23531	0,4563	0,4281	0,97705	0,18482	0,12142
0,125	0,35355	0,3536	0,3398	0,27657	0,4869	0,4531	0,96909	0,21722	0,14875
0,150	0,38731	0,3873	0,3695	0,31478	0,5145	0,4751	0,96094	0,24723	0,17976
0,175	0,41833	0,4183	0,3962	0,35047	0,5398	0,4949	0,95278	0,27526	0,20497
0,200	0,44721	0,4472	0,4205	0,38402	0,5633	0,5132	0,94474	0,3061	0,23488
0,225	0,47434	0,4743	0,4429	0,41573	0,5854	0,5296	0,93691	0,32651	0,26133
0,250	0,50000	0,5000	0,4636	0,44583	0,6062	0,5451	0,92938	0,35015	0,28804
0,275	0,52441	0,5244	0.4831	0,47449	0,6261	0,5593	0,92222	0,37266	0,31305
0,300	0,54772	0,5477	0,5011	0,50187	0,6449	0,5728	0,91549	0,39417	0,33751
0,325	0,57009	0,5701	0,5181	0,52809	0,6631	0,5855	0,90926	0,41476	0,36096
0,350	0,59161	0,5916	0,5342	0,55326	0,6804	0,5975	0,90355	0,43453	0,38603
0,375	0,61237	0,6124	0,5495	0,57746	0,6972	0,6088	0,89842	0,45354	0,40712
0,400	0,63246	0,6325	0,5639	0,60078	0,7133	0,6196	0,89391	0,47185	0,42882
0,425	0,65192	0,6519	0,5777	0,62328	0,7291	0,6199	0,89002	0,48952	0,44841
0,450	0,67082	0,6708	0,5909	0,64503	0,7441	0,6397	0,88682	0,50661	0,46852
0,475	0,68921	0,6892	0,6034	0,66606	0,7588	0,6491	0,88431	0,52312	0,48931
0,500	0,70711	0,7071	0,6155	0,68644	0,7731	0,6581	0,88252	0,53913	0,50923
0,525	0,72457	0,7246	0,6271	0,70622	0,7872	0,6667	0,88142	0,55465	0,52702
0,550	0,74162	0,7416	0,6381	0,72538	0,8005	0,6751	0,88109	0,56972	0,54504
0,575	0,75829	0,7583	0,6488	0,74402	0,8137	0,6832	0,88151	0,58435	0,56381
0,600	0,77462	0,7746	0,6591	0,76215	0,8265	0,6907	0,88268	0,59859	0,57995
0,625	0,79057	0,7906	0,6691	0,77979	0,8391	0,6981	0,88462	0,61245	0,59802
0,650	0,80623	0,8062	0,6785	0,79697	0,8514	0,7053	0,88732	0,62594	0,61303
0,675	0,82158	0,8216	0,6878	0,81372	0,8634	0,7122	0,89078	0,63909	0,62741
0,700	0,83666	0,8367	0,6967	0,83006	0,8751	0,7189	0,89499	0,65192	0,64176
0,725	0,85147	0,8515	0,7050	0,84612	0,8866	0,7254	0,89995	0,66445	0,65531
0,750	0,86603	0,8660	0,7137	0,86157	0,8979	0,7317	0,90565	0,67667	0,67003
0,775	0,88034	0,8803	0,7218	0,87678	0,9091	0,7377	0,91208	0,68862	0,68331
0,800	0,89443	0,8944	0,7297	0,89165	0,9198	0,7437	0,91922	0,70031	0,69611
0,825	0,90832	0,9083	0,7374	0,90622	0,9305	0,7494	0,92705	0,71173	0,70702
0,850	0,92195	0,9221	0,7448	0,92043	0,9409	0,7552	0,93557	0,72291	0,72021
0,875	0,93541	0,9354	0,7521	0,93437	0,9512	0,7604	0,94476	0,73385	0,73111
0,900	0,94868	0,9487	0,7591	0,94803	0,9613	0,7657	0,95459	0,74458	0,74336
0,925	0,96177	0,9618	0,7659	0,96142	0,9712	0,7708	0,96505	0,75508	0,75402
0,950	0,97468	0,9747	0,7726	0,97452	0,9809	0,7758	0,97612	0,76539	0,76512
0,975	0,98742	0,9874	0,7791	0,98738	0,9906	0,7807	0,98778	0,77549	0,77503
1	1	1	0,7854	1	1	0,7854	1	0,78539	0,78578

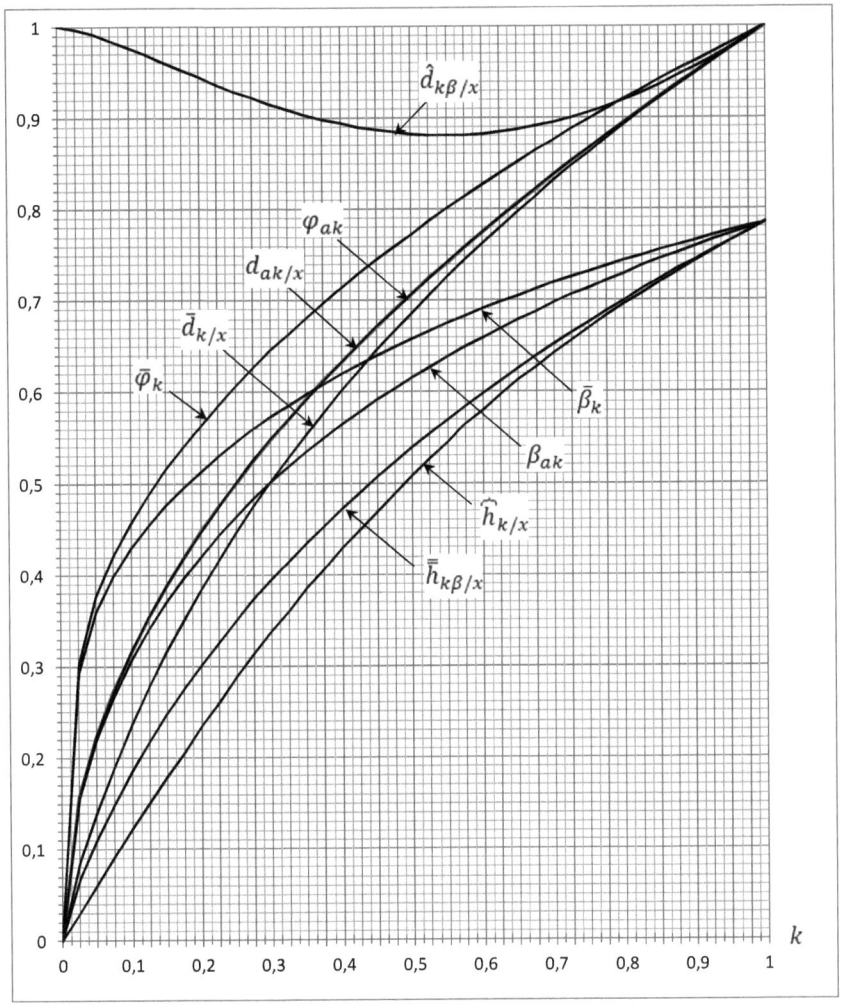

Fig. 7. Graphs of the functions diameters, angular coefficients and angles of rotation, for example, quasi-endless set of parallel chords of an ellipse.

The *Table 3* and *Fig. 8* are complementary to the *Table 2* and *Fig. 7*.

Table 3

k	$\hat{\varphi}_k$	$\hat{\beta}_k$	$\bar{\beta}_k^o$	β_{ak}^o	$\hat{\beta}_k^o$	k	$\hat{\varphi}_k$	$\hat{\beta}_k$	$\bar{\beta}_k^o$	β_{ak}^o	$\hat{\beta}_k^o$
0	0	0	0	0	0						
0,025	-0,002	-0	16,83	8,985	-0,12	0,525	-0,35	-0,34	38,20	35,93	-19,3
0,050	-0,007	-0,01	20,63	12,60	-0,38	0,550	-0,38	-0,36	38,68	36,56	-20,7
0,075	-0,013	-0,01	22,86	15,32	-0,76	0,575	-0,41	-0,39	39,13	37,17	-22,1
0,100	-0,02	-0,02	24,53	17,55	-1,26	0,600	-0,44	-0,41	39,57	37,76	-23,5
0,125	-0,03	-0,03	25,96	19,47	-1,84	0,625	-0,47	-0,44	40,00	38,33	-25,0
0,150	-0,04	-0,04	27,22	21,17	-2,50	0,650	-0,50	-0,46	40,41	38,88	-26,4
0,175	-0,06	-0,06	28,36	22,71	-3,25	0,675	-0,53	-0,49	40,81	39,41	-27,8
0,200	-0,07	-0,07	29,39	22,09	-4,06	0,700	-0,56	-0,51	41,19	39,92	-29,2
0,225	-0,09	-0,09	30,34	25,38	-4,94	0,725	-0,59	-0,54	41,56	40,41	-30,7
0,250	-0,10	-0,10	31,23	26,57	-5,89	0,750	-0,63	-0,56	41,92	40,89	-32,1
0,275	-0,12	-0,12	32,05	27,67	-6,89	0,775	-0,66	-0,58	42,27	41,36	-33,5
0,300	-0,14	-0,14	32,82	28,71	-7,94	0,800	-0,70	-0,61	42,61	41,81	-34,8
0,325	-0,16	-0,16	33,55	29,69	-9,05	0,825	-0,73	-0,63	42,94	42,25	-36,2
0,350	-0,18	-0,18	34,23	30,61	-10,2	0,850	-0,77	-0,65	43,26	42,67	-37,5
0,375	-0,20	-0,20	34,88	31,48	-11,4	0,875	-0,81	-0,68	43,57	43,09	-38,8
0,400	-0,22	-0,22	35,50	32,31	-12,6	0,900	-0,84	-0,70	43,87	43,49	-40,1
0,425	-0,25	-0,24	36,09	33,10	-13,9	0,925	-0,88	-0,72	44,16	43,88	-41,4
0,450	-0,27	-0,27	36,65	33,85	-15,2	0,950	-0,92	-0,74	44,45	44,27	-42,6
0,475	-0,30	-0,29	37,19	34,57	-16,6	0,975	-0,96	-0,76	44,73	44,64	-43,8
0,500	-0,32	-0,31	37,71	35,26	-17,9	1,000	-1,00	-0,79	45,00	45,00	-45,0

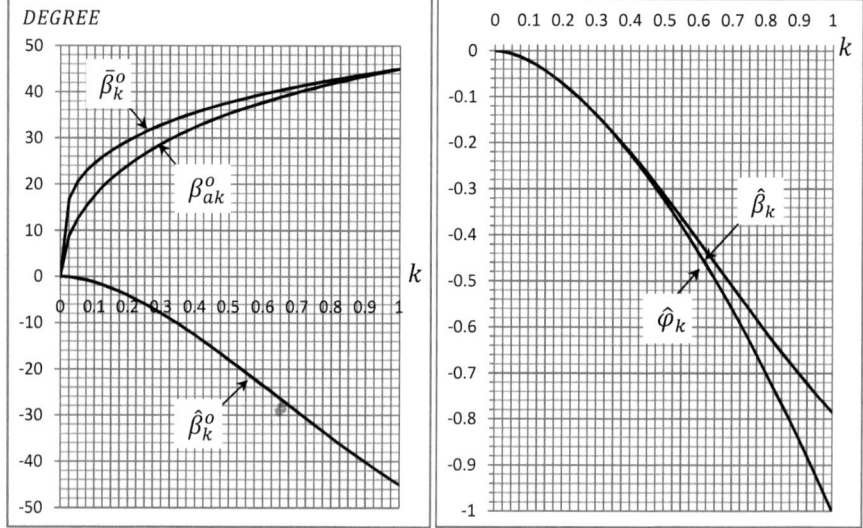

Fig.8. The dependence of the slope and the angles of rotation of diameters, depending on the compression ratio (k) of the ellipse, where:

$\bar{\beta}_k^o = f(k)$ − one of the two angles of rotation ($\pm\bar{\beta}_k^o$) relative average diameter ($\bar{d}_{k/x}$) of the ellipse, in degrees;

$\beta_{ak}^o = f(k)$ − one of the two angles of rotation ($\pm\beta_{ak}^o$) relative equivalent diameter ($d_{ak/x}$) of the ellipse, in degrees;

$\hat{\varphi}_k = f(k)$ and $\hat{\beta}_k = f(k)$ − one of the two slopes ($\pm\hat{\varphi}_k$) and rotation angles ($\pm\hat{\beta}_k$) of the relative diameter ($\hat{d}_{k/x}$) the conjugate set of chords, parallel to the relative average diameter ($\bar{d}_{k/x}$) of the ellipse, in radians;

$\hat{\beta}_k^o = f(k)$ − one of the two rotation angles (($\pm\hat{\beta}_k$), in degrees.

4. Ratio of the average chord sets of parallel chords of an ellipse to the diameter of this set (from simulation results)

According to the results of computer simulation of chords and diameters of ellipses defined ratios average chord ($\bar{h}_{\beta/x}$) and diameter ($d_{\beta/x}$) in each set $Q\{h_{\beta i/x}\}$ of parallel chords. Based on these results [2] concluded that at least diminishing definite limit relative step between the chords ($\Delta\vec{h}_x \to 0$), the ratio ($\bar{h}_{\beta/x}/d_{\beta/x}$) → $\pi/4$. Already at ($\Delta\vec{h}_x = 0,0005$), deviation ratio from the limit ($\pi/4$), to which it is tends, is about $0,01\%$. This pattern is expressed by the equation

$$\lim_{\Delta h_x \to 0}(\bar{h}_{\beta/x}/d_{\beta/x}) \approx \pi/4, \quad \text{or} \quad \bar{h}_{\beta/x,(\Delta h \to 0)}/d_{\beta/x} = \pi/4. \quad (19)$$

Thus, it is concluded that for infinite sets $Q\{h_{\beta i/x}\}_{\Delta h_x \to 0}$ of parallel chords with any rotation angle (β), the ratio any of the two averages chords ($\bar{h}_{\beta/x}$) to the equal direction diameter ($d_{\beta/x}$) in any ellipse, as in the circle ($k = 1$), is constant, equal to ($\pi/4$).

On the basis of (2) and (19), a formula for calculating the average relative chord ($\bar{h}_{\beta/x}$) an infinite set of parallel chords $Q\{h_{\beta i/x}\}_{\Delta h_x \to 0}$ of the ellipse, which is the current rotation angle (β)

$$\bar{h}_{\beta/x,(\Delta h \to 0)} = \frac{\pi}{4}\, d_{\beta/x} = \frac{\pi}{4}k\sqrt{\frac{1+tan^2\beta_i}{k^2+tan^2\beta_i}} = \frac{\pi}{4}k\sqrt{\frac{1+\varphi_{\beta i}^2}{k^2+\varphi_{\beta i}^2}}. \quad (20)$$

4.1. Ratio of the average chord sets of parallel chords of an ellipse to the diameter of this set (proof)

To check and proof of the correctness of the results obtained by computer simulation [2], considered known [1^{451}] Theorem on the mean of integral calculus. According to the theorem, the definite integral of the first quarter of the ellipse is equal to the product of *Fig. 6* the length of the integration interval (0, a) the value of the integrand at some point (ξ) of the interval (0, a), which is average for this interval

$$\int_0^a f(x)dx = (a - 0)f(\xi).$$

This integral is the area [1^{483}] curvilinear trapezoid, representing the first quarter of the ellipse (0, $25S_k$) with a compression ratio (k).

It is also known [1^{485}], that the fourth part of the area (S_k) ellipse is
$0,25\ S_k = 0,25\ \pi ab$, where a and b semiaxis of the ellipse.

By comparing the two expressions for the same area, we have: $f(\xi) = 0,25\pi b$ and the ratio of $f(\xi)/b = \pi/4$ is the ratio of the average semichord ($\bar{h}_y/2 = a$) parallel to her principal semidiameter ($d_y/2 = b$) of the ellipse. Similar expressions are obtained by turns the ellipse at an angle ($\pm \pi / 2$):

$$\frac{\bar{h}_y}{d_y} = \frac{\bar{h}_x}{d_x} = \frac{\pi}{4}.$$

The relations obtained are sufficient proof of the correctness of the expressions (19, 20).

5. Subellipse averages chords

In any ellipse can construct an infinite set of parallel chords and an infinite set, corresponding, average chords ($\bar{h}_{\beta(\Delta h \to 0)}$), which together fill the area, whose shape is similar to the shape of the ellipse considered.

Assign the resulting figure called " *Subellipse averages chords* " (*Fig. 9*).

5.1. Comparison of diameters of the ellipse and the subellipse

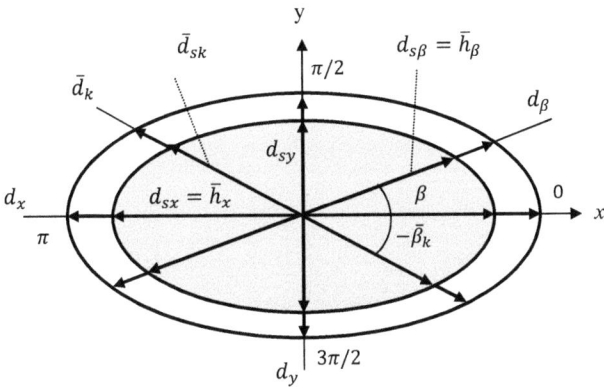

Fig. 9. Subellipse averages chords, formed by the rotation of the average chord, rotating variable sets of parallel chords of primary ellipse, where:

$d_{sx} = \bar{h}_x$ and $d_{sy} = \bar{h}_y$ — major diameters of the subellipse, equal to the arithmetic average chord of the equal directed set of parallel chords of primary ellipse;

$d_{s\beta} = \bar{h}_\beta$ — variable diameter with the current rotation angle (β), for circular rotation which forms an subellipse, and equal to the arithmetic average chord, rotating variable set of parallel chords of primary ellipse;

$\bar{d}_{sk}$ — one of the two arithmetic average diameters of the subellipse, which is the angle of rotation ($\pm\bar{\beta}_k$);

$\bar{d}_k$ — one of the two primary arithmetic average diameters of the primary ellipse, which is the angle of rotation ($\pm\bar{\beta}_k$).

The ratio of the same name diameters between of the subellipse and of the primary ellipse

$$\frac{d_{s\beta}}{d_\beta} = \frac{\bar{d}_{sk}}{\bar{d}_k} = \frac{d_{sx}}{d_x} = \frac{d_{sy}}{d_y} = \frac{\pi}{4} \tag{21}$$

Ratio area of the subellipse averages chords $S_{sk} = (\pi/4)^3 k d_x^2$ to area of the primary ellipse $S_k = (\pi/4)k d_x^2$

$$\frac{S_{sk}}{S_k} = \left(\frac{\pi}{4}\right)^2 \tag{22}$$

5.2. Comparison of the relative areas of different figures

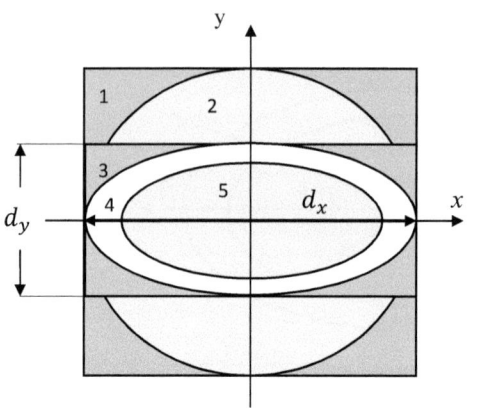

For comparison, in *Fig. 10* presented ranked in order of ative areas $(S_{k/x^2} = S_k/d_x^2)$ various figures.

Fig. 10. Comparison of the relative areas of different figures, obtained by dividing the area of figures on (d_x^2), where:

1 – Square, $\qquad\qquad\qquad\qquad\qquad S_{1/x^2} = 1$

2 – Circle $\qquad\qquad\qquad\qquad\qquad\quad S_{2/x^2} = \left(\frac{\pi}{4}\right);$

3 – Rectangle, for $k = d_y/d_x = [0; 1]$: $\qquad S_{3/x^2} = [0; k];$

4 – Ellipse, for $k = [0; 1]$: $\qquad\qquad\quad S_{4/x^2} = [0; \left(\frac{\pi}{4}\right)k];$

5 – Subellipse averages chords, for $k = [0; 1]$: $\qquad S_{5/x^2} = [0; \left(\frac{\pi}{4}\right)^3 k].$

6. Average chord of the ellipse

Subellipse infinite set of average chords in shape similar to primary (parent) ellipse, their compression ratios are equal. Vectors of the two average diameters $(\bar{d}_{sk})$ subellipse (*Fig. 9*) coincide with vectors pairs of the arithmetic average values $(\bar{\bar{h}}_{k\beta/x(\Delta h \to 0)})$ of the average relative chords, step though rotating infinite sets of the primary ellipse.

For infinite sets of parallel chords, the ratio of the arithmetic average $(\bar{\bar{h}}_{k\beta/x(\Delta h \to 0)})$ of the averages relative chords to the average diameter $(\bar{d}_{k/x})$ primary ellipse (21) is equal to

$$\lim_{\Delta h \to 0} \bar{d}_{sk/x}/\bar{d}_{k/x} = \lim_{\Delta h \to 0}(\bar{\bar{h}}_{k\beta/x}/\bar{d}_{k/x}) = \bar{\bar{h}}_{k\beta/x(\Delta h \to 0)}/\bar{d}_{k/x} = \boldsymbol{\pi/4}. \qquad (23)$$

With the expressions (3) and (23) we obtain

$$\bar{\bar{h}}_{k\beta/x(\Delta h \to 0)} = \frac{\pi}{4}\,\bar{d}_{k/x} = \frac{\pi}{4}\,k\,\sqrt{\frac{1+tan^2\bar{\beta}_k}{k^2+tan^2\bar{\beta}_k}} = \frac{\pi}{4}k\,\sqrt{\frac{1+\bar{\varphi}_k{}^2}{k^2+\bar{\varphi}_k{}^2}};\qquad(24)$$

Of standard statistical parameters obtained during the simulation [2], when dealing with the same quasi-endless array of chords, there is a mismatch results of calculation values $(\bar{\bar{h}}_{k\beta/x(\Delta\bar{h}_x)})$ and the arithmetic average values $(\overset{\frown}{h}_{k/x,\Delta\bar{h}_x})$ aggregate set $(\tilde{Q}\{h_{k\beta i/x}\}_{\Delta\bar{h}_x})$ all chords of the ellipse $(\bar{\bar{h}}_{k\beta/x} \geq \overset{\frown}{h}_{k/x})$, hereinafter referred to as: the arithmetic average of the statistical set of relative chords of an ellipse. An exception is the case, when $k = 1$ (circle).

Thus, for both infinite and finite, rotating variable set $(\tilde{Q}\{h_{k\beta i/x}\}_{\Delta\bar{h}_x})$ parallel chords of an ellipse, you need to distinguish between the two results averaging the entire array of chords of the ellipse:

$\bar{\bar{h}}_{k\beta/x}$ — arithmetic average of average chord, rotating variable set of parallel relative chords of an ellipse, whose the compression ratio (k), or the average relative diameter of the subellipse averages chords;

$\overset{\frown}{h}_{k/x}$ — arithmetic average of the statistical set of relative chords of an ellipse.

The results of calculations [2] averaged relative chords $(\bar{\bar{h}}_{k\beta/x}, \overset{\frown}{h}_{k/x})$ for quasi-endless sets of ellipses, in some cases, due to immaterial errors can be regarded as the results for infinite sets. After interpolation data on range compression ratio $k = [0, 1]$, the results presented in *Table 2* and in *Figure 7* as graphs of functions $\bar{\bar{h}}_{k\beta/x} = f(k)$, $\overset{\frown}{h}_k = f(k)$.

Similar results for finite sets of parallel relative chords of ellipses, as a sample of individual tables, are presented in *Appendix*.

7. Conclusion

1. All the above parameters are presented on a full interval compression ratio $k = [0, 1]$ ellipses in relative units (RLU) obtained by dividing the actual value of the parameter on the big major diameter of the ellipse (d_x). Transition to real parameters in reverse order.

2. By computer simulation, step through rotating variable set ($\tilde{Q}\{h_{\beta i/x}\}, d_\beta \in \tilde{Q}$) relative parallel chords of an ellipse with the specified relative step ($\Delta\vec{h}_x$) between chords, obtained a series of stepwise sets of parallel relative chords ($Q\{h_{\beta i/x}\}_{\Delta\vec{h}_x}$). Within a single ellipse, chord of all stepwise sets to produce an aggregate array ($\tilde{Q}\{h_{k\beta i/x}\}_{\Delta\vec{h}_x}$) chords of the ellipse.

3. For ellipses with the current value (k_i), determined the relative values of the average chord ($\bar{h}_{\beta i/x}$) for each step through set $Q\{h_{\beta i/x}\}$ and the arithmetic average ($\bar{\bar{h}}_{k\beta/x}$) of these averages for the entire ellipse. In parallel, is defined the arithmetic average ($\vec{\tilde{h}}_{k/x}$) aggregate array ($\tilde{Q}\{h_{k\beta i/x}\}_{\Delta\vec{h}_x}$) the relative chord of the ellipse, designated as the arithmetic average of the statistical set of chords of an ellipse.

4. The interval $k = [0, 1]$, the functions averaged chords ($\bar{\bar{h}}_{k\beta/x}$) and ($\vec{\tilde{h}}_{k/x}$) of the ellipse are nonlinear, their current values ($\bar{\bar{h}}_{k\beta/x} \geq \vec{\tilde{h}}_k$) coincide only in the special case where $k_i = 1$ (circle).

5. In step through sets of parallel chords (including diameters) are defined relative diameters ($d_{\beta i/x}$) and their arithmetic average ($\bar{d}_{k/x}$) for the entire ellipse, simply call the average diameter of the ellipse.

6. For ellipses with the current value (k_i) ratio of the average chord ($\bar{h}_{\beta/x(\Delta h \to 0)}$), in each step through rotating infinite set ($Q\{h_{\beta i/x}\}_{\Delta h_x \to 0}$) of parallel chords of an ellipse to the co-rotational diameter ($d_{\beta i/x}$), for any current rotation angle (β) sets, as well ($\pi / 4$). The ratio of the arithmetic average ($\bar{\bar{h}}_{k\beta/x(\Delta h \to 0)}$) of average

chord ($\bar{\bar{h}}_{\beta/x(\Delta h \to 0)}$), a rotating variable infinite set of parallel chords of an ellipse to the average diameter ($\bar{d}_{k/x}$) of the ellipse is also independent size and shape of the ellipse, and also for circle is constant:

$$\bar{\bar{h}}_{k\beta/x(\Delta h \to 0)}/\bar{d}_{k/x} = \bar{h}_{\beta/x(\Delta h \to 0)}/d_{\beta i/x} = \pi/4 = const.$$

7. The totality of an infinite set of the average chords ($\bar{h}_{\beta/x(\Delta h \to 0)}$) of the ellipse forms the *subellipse averages chords*, in shape similar the primary ellipse, and the area which in $(\frac{\pi}{4})^2$ times less of its area. Vectors of the two average diameters subellipse coincide with pairs of arithmetic average values of average chord ($\bar{\bar{h}}_{k\beta/x(\Delta h \to 0)}$) of the primary ellipse.

8. Limit values of the sum of set parallel relative chords of a single ellipse ($\sum_{\Delta \vec{h}_x} H_{\beta i/x}$), for specified step ($\Delta \vec{h}_x$) between the chords, are equal in all directions and depend from the specified limit distance ($\Delta \vec{h}_x$) between chords and compression ratio (k_i) of the ellipse.

9. The average diameter ($\bar{d}_{k/x}$) is not equal to the equivalent ($d_{ak/x}$) diameter of the ellipse, it is less than it. Are not equal to their angular coefficients (slopes) and rotation angles.

10. Received universal graphs of nonlinear functions, allow you to define ellipses for any shape and size with the same name values:

$\bar{\bar{h}}_{k\beta/x} = f(k)$ – arithmetic average of average chord, rotating variable set of parallel relative chords of an ellipse;

$\hat{h}_{k/x} = f(k)$ – arithmetic average of the statistical set of relative chords of an ellipse;

$\bar{d}_{k/x} = f(k)$ – average relative diameter of the ellipse;

$d_{ak/x} = f(k)$ – equivalent relative diameter of the ellipse;

$\hat{d}_{k/x} = f(k)$ – relative diameter, conjugate to set of parallel chords, including the average relative diameter ($\bar{d}_{k/x}$) of the ellipse;

$\bar{\varphi}_k = f(k)$, $\varphi_{ak} = f(k)$, $\hat{\varphi}_k = f(k)$, $\bar{\beta}_k = f(k)$, $\beta_{ak} = f(k)$, $\hat{\beta}_k = f(k)$ – angular coefficients (slopes) and rotation angles corresponding diameters of the ellipse in radians, for a pair of average diameter ($\pm\bar{\varphi}_k$, $\pm\bar{\beta}_k$) shows only positive values;

$\bar{\beta}_k^o = f(k)$, $\beta_{ak}^o = f(k)$, $\hat{\beta}_k^o = f(k)$ – rotation angles corresponding diameters of the ellipse in degrees.

Bibliography

[1]   М.Я. Выгодский. Справочник по высшей математике. М, 1998., 864с. *Mathematical Handbook: Higher Mathematics by M. Vygodsky, G. Yankovsky (Translator)*.2011.

[2]   www.annov.de/ellipse/Calculation_table_of_chords_and_diameters_of_the_ellipses.xlsx (1 Mb)

Appendix. Sampling calculation of tables for finite sets. $k = 0,9$

k \ β	0,17	0,35	0,52	0,7	0,87	1,05	1,22	1,4	1,57	1,75	1,92	2,09	2,27	2,44	2,62	2,79	2,97	3,14
0,9 Δh/a																		
1																		
0,99								0,11	0,13	0,11								
0,98							0,12	0,17	0,18	0,17	0,12							
0,97						0,1	0,18	0,21	0,22	0,21	0,18	0,1						
0,96						0,17	0,22	0,24	0,25	0,24	0,22	0,17						
0,95					0,13	0,21	0,25	0,27	0,28	0,27	0,25	0,21	0,13					
0,94				0,07	0,19	0,25	0,28	0,3	0,31	0,3	0,28	0,25	0,19	0,07				
0,93				0,16	0,23	0,28	0,31	0,33	0,33	0,33	0,31	0,28	0,23	0,16				
0,92			0,11	0,21	0,27	0,31	0,33	0,35	0,35	0,35	0,33	0,31	0,27	0,21	0,11			
0,91		0,07	0,18	0,25	0,3	0,33	0,36	0,37	0,37	0,37	0,36	0,33	0,3	0,25	0,18	0,07		
0,9	0,08	0,16	0,23	0,28	0,33	0,36	0,38	0,39	0,39	0,39	0,38	0,36	0,33	0,28	0,23	0,16	0,08	
0,89	0,17	0,22	0,27	0,31	0,35	0,38	0,4	0,41	0,41	0,41	0,4	0,38	0,35	0,31	0,27	0,22	0,17	0,15
0,88	0,22	0,26	0,3	0,34	0,37	0,4	0,42	0,42	0,43	0,42	0,42	0,4	0,37	0,34	0,3	0,26	0,22	0,21
0,87	0,27	0,3	0,33	0,37	0,4	0,42	0,43	0,44	0,44	0,44	0,43	0,42	0,4	0,37	0,33	0,3	0,27	0,26
0,86	0,3	0,33	0,36	0,39	0,42	0,44	0,45	0,46	0,46	0,46	0,45	0,44	0,42	0,39	0,36	0,33	0,3	0,29
0,85	0,34	0,36	0,39	0,41	0,44	0,45	0,47	0,47	0,47	0,47	0,47	0,45	0,44	0,41	0,39	0,36	0,34	0,33
0,84	0,37	0,38	0,41	0,43	0,45	0,47	0,48	0,49	0,49	0,49	0,48	0,47	0,45	0,43	0,41	0,38	0,37	0,36
0,83	0,39	0,41	0,43	0,45	0,47	0,49	0,49	0,5	0,5	0,5	0,49	0,49	0,47	0,45	0,43	0,41	0,39	0,39
0,82	0,42	0,43	0,45	0,47	0,49	0,5	0,51	0,51	0,52	0,51	0,51	0,5	0,49	0,47	0,45	0,43	0,42	0,41
0,81	0,44	0,45	0,47	0,49	0,5	0,51	0,52	0,53	0,53	0,53	0,52	0,51	0,5	0,49	0,47	0,45	0,44	0,44
0,8	0,46	0,47	0,49	0,5	0,52	0,53	0,53	0,54	0,54	0,54	0,53	0,53	0,52	0,5	0,49	0,47	0,46	0,46
0,79	0,48	0,49	0,51	0,52	0,53	0,54	0,55	0,55	0,55	0,55	0,55	0,54	0,53	0,52	0,51	0,49	0,48	0,48
0,78	0,5	0,51	0,52	0,54	0,55	0,55	0,56	0,56	0,56	0,56	0,56	0,55	0,55	0,54	0,52	0,51	0,5	0,5
0,77	0,52	0,53	0,54	0,55	0,56	0,57	0,57	0,57	0,57	0,57	0,57	0,57	0,56	0,55	0,54	0,53	0,52	0,52
0,76	0,54	0,55	0,56	0,56	0,57	0,58	0,58	0,58	0,58	0,58	0,58	0,58	0,57	0,56	0,56	0,55	0,54	0,54
0,75	0,56	0,56	0,57	0,58	0,59	0,59	0,59	0,59	0,6	0,59	0,59	0,59	0,59	0,58	0,57	0,56	0,56	0,55
0,74	0,57	0,58	0,58	0,59	0,6	0,6	0,6	0,6	0,61	0,6	0,6	0,6	0,6	0,59	0,58	0,58	0,57	0,57
0,73	0,59	0,59	0,6	0,6	0,61	0,61	0,61	0,61	0,62	0,61	0,61	0,61	0,61	0,6	0,6	0,59	0,59	0,58
0,72	0,6	0,61	0,61	0,62	0,62	0,62	0,62	0,62	0,62	0,62	0,62	0,62	0,62	0,62	0,61	0,61	0,6	0,6
0,71	0,62	0,62	0,62	0,63	0,63	0,63	0,63	0,63	0,63	0,63	0,63	0,63	0,63	0,63	0,62	0,62	0,62	0,61
0,7	0,63	0,63	0,64	0,64	0,64	0,64	0,64	0,64	0,64	0,64	0,64	0,64	0,64	0,64	0,64	0,63	0,63	0,63
0,69	0,64	0,65	0,65	0,65	0,65	0,65	0,65	0,65	0,65	0,65	0,65	0,65	0,65	0,65	0,65	0,65	0,64	0,64
0,68	0,66	0,66	0,66	0,66	0,66	0,66	0,66	0,66	0,66	0,66	0,66	0,66	0,66	0,66	0,66	0,66	0,66	0,66
0,67	0,67	0,67	0,67	0,67	0,67	0,67	0,67	0,67	0,67	0,67	0,67	0,67	0,67	0,67	0,67	0,67	0,67	0,67
0,66	0,68	0,68	0,68	0,68	0,68	0,68	0,68	0,68	0,68	0,68	0,68	0,68	0,68	0,68	0,68	0,68	0,68	0,68
0,65	0,69	0,69	0,69	0,69	0,69	0,69	0,69	0,68	0,68	0,68	0,69	0,69	0,69	0,69	0,69	0,69	0,69	0,69
0,64	0,7	0,7	0,7	0,7	0,7	0,7	0,69	0,69	0,69	0,69	0,69	0,7	0,7	0,7	0,7	0,7	0,7	0,7
0,63	0,71	0,71	0,71	0,71	0,71	0,7	0,7	0,7	0,7	0,7	0,7	0,7	0,71	0,71	0,71	0,71	0,71	0,71
0,62	0,72	0,72	0,72	0,72	0,72	0,71	0,71	0,71	0,71	0,71	0,71	0,71	0,72	0,72	0,72	0,72	0,72	0,72
0,61	0,73	0,73	0,73	0,73	0,72	0,72	0,72	0,71	0,71	0,71	0,72	0,72	0,72	0,73	0,73	0,73	0,73	0,74
0,6	0,74	0,74	0,74	0,74	0,73	0,73	0,72	0,72	0,72	0,72	0,72	0,73	0,73	0,74	0,74	0,74	0,74	0,75
0,59	0,75	0,75	0,75	0,74	0,74	0,73	0,73	0,73	0,73	0,73	0,73	0,73	0,74	0,74	0,75	0,75	0,75	0,76
0,58	0,76	0,76	0,76	0,75	0,75	0,74	0,74	0,73	0,73	0,73	0,74	0,74	0,75	0,75	0,76	0,76	0,76	0,76
0,57	0,77	0,77	0,77	0,76	0,75	0,75	0,74	0,74	0,74	0,74	0,74	0,75	0,75	0,76	0,77	0,77	0,77	0,77
0,56	0,78	0,78	0,77	0,77	0,76	0,76	0,75	0,75	0,75	0,75	0,75	0,76	0,76	0,77	0,77	0,78	0,78	0,78
0,55	0,79	0,79	0,78	0,78	0,77	0,76	0,76	0,75	0,75	0,75	0,76	0,76	0,77	0,78	0,78	0,79	0,79	0,79
0,54	0,8	0,8	0,79	0,78	0,78	0,77	0,76	0,76	0,76	0,76	0,76	0,77	0,78	0,78	0,79	0,8	0,8	0,8
0,53	0,81	0,8	0,8	0,79	0,78	0,77	0,77	0,76	0,76	0,76	0,77	0,77	0,78	0,79	0,8	0,8	0,81	0,81
0,52	0,81	0,81	0,8	0,8	0,79	0,78	0,77	0,77	0,77	0,77	0,77	0,78	0,79	0,8	0,8	0,81	0,81	0,82
0,51	0,82	0,82	0,81	0,8	0,79	0,78	0,78	0,77	0,78	0,78	0,78	0,79	0,79	0,8	0,81	0,82	0,82	0,82
0,5	0,83	0,83	0,82	0,81	0,8	0,79	0,79	0,78	0,78	0,78	0,79	0,79	0,8	0,81	0,82	0,83	0,83	0,83
0,49	0,84	0,83	0,82	0,82	0,81	0,8	0,79	0,79	0,78	0,79	0,79	0,8	0,81	0,82	0,82	0,83	0,84	0,84
0,48	0,84	0,84	0,83	0,82	0,81	0,8	0,8	0,79	0,79	0,79	0,8	0,8	0,81	0,82	0,83	0,84	0,84	0,85
0,47	0,85	0,85	0,84	0,83	0,82	0,81	0,8	0,8	0,79	0,8	0,8	0,81	0,82	0,83	0,84	0,85	0,85	0,85
0,46	0,86	0,85	0,84	0,83	0,82	0,81	0,81	0,8	0,8	0,8	0,81	0,81	0,82	0,83	0,84	0,85	0,86	0,86
0,45	0,86	0,86	0,85	0,84	0,83	0,82	0,81	0,81	0,8	0,8	0,81	0,82	0,83	0,84	0,85	0,86	0,86	0,87
0,44	0,87	0,86	0,86	0,84	0,83	0,82	0,82	0,81	0,81	0,81	0,82	0,82	0,83	0,84	0,86	0,86	0,87	0,87
0,43	0,88	0,87	0,86	0,85	0,84	0,83	0,82	0,81	0,81	0,81	0,82	0,83	0,84	0,85	0,86	0,87	0,88	0,88
0,42	0,88	0,88	0,87	0,85	0,84	0,83	0,82	0,82	0,82	0,82	0,82	0,83	0,84	0,85	0,87	0,88	0,88	0,88
0,41	0,89	0,88	0,87	0,86	0,85	0,84	0,83	0,82	0,82	0,82	0,83	0,84	0,85	0,86	0,87	0,88	0,89	0,89
0,4	0,89	0,89	0,88	0,86	0,85	0,84	0,83	0,83	0,82	0,83	0,83	0,84	0,85	0,86	0,88	0,89	0,89	0,9
0,39	0,9	0,89	0,88	0,87	0,86	0,85	0,84	0,83	0,83	0,83	0,84	0,85	0,86	0,87	0,88	0,89	0,9	0,9
0,38	0,9	0,9	0,89	0,87	0,86	0,85	0,84	0,83	0,83	0,83	0,84	0,85	0,86	0,87	0,89	0,9	0,9	0,91
0,37	0,91	0,9	0,89	0,88	0,87	0,85	0,84	0,84	0,84	0,84	0,84	0,85	0,87	0,88	0,89	0,9	0,91	0,91
0,36	0,91	0,91	0,9	0,88	0,87	0,86	0,85	0,84	0,84	0,84	0,85	0,86	0,87	0,88	0,9	0,91	0,91	0,92
0,35	0,92	0,91	0,9	0,89	0,87	0,86	0,85	0,85	0,84	0,85	0,85	0,86	0,87	0,89	0,9	0,91	0,92	0,92
0,34	0,92	0,92	0,9	0,89	0,88	0,86	0,85	0,85	0,85	0,85	0,85	0,86	0,88	0,89	0,9	0,92	0,92	0,93

0,33	0,93	0,92	0,91	0,89	0,88	0,87	0,86	0,85	0,85	0,85	0,86	0,87	0,88	0,89	0,91	0,92	0,93	0,93
0,32	0,93	0,92	0,91	0,9	0,88	0,87	0,86	0,85	0,85	0,85	0,86	0,87	0,88	0,9	0,91	0,92	0,93	0,93
0,31	0,94	0,93	0,92	0,9	0,89	0,87	0,86	0,86	0,86	0,86	0,86	0,87	0,89	0,9	0,92	0,93	0,94	0,94
0,3	0,94	0,93	0,92	0,91	0,89	0,88	0,87	0,86	0,86	0,86	0,87	0,88	0,89	0,91	0,92	0,93	0,94	0,94
0,29	0,94	0,94	0,92	0,91	0,89	0,88	0,87	0,86	0,86	0,86	0,87	0,88	0,89	0,91	0,92	0,94	0,94	0,95
0,28	0,95	0,94	0,93	0,91	0,9	0,88	0,87	0,87	0,86	0,87	0,87	0,88	0,9	0,91	0,93	0,94	0,95	0,95
0,27	0,95	0,94	0,93	0,91	0,9	0,89	0,88	0,87	0,87	0,87	0,88	0,89	0,9	0,91	0,93	0,94	0,95	0,95
0,26	0,95	0,95	0,93	0,92	0,9	0,89	0,88	0,87	0,87	0,87	0,88	0,89	0,9	0,92	0,93	0,95	0,95	0,96
0,25	0,96	0,95	0,94	0,92	0,91	0,89	0,88	0,87	0,87	0,87	0,88	0,89	0,91	0,92	0,94	0,95	0,96	0,96
0,24	0,96	0,95	0,94	0,92	0,91	0,89	0,88	0,88	0,87	0,88	0,88	0,89	0,91	0,92	0,94	0,95	0,96	0,96
0,23	0,96	0,95	0,94	0,93	0,91	0,9	0,89	0,88	0,88	0,88	0,89	0,9	0,91	0,93	0,94	0,95	0,96	0,97
0,22	0,97	0,96	0,94	0,93	0,91	0,9	0,89	0,88	0,88	0,88	0,89	0,9	0,91	0,93	0,94	0,96	0,97	0,97
0,21	0,97	0,96	0,95	0,93	0,91	0,9	0,89	0,88	0,88	0,88	0,89	0,9	0,91	0,93	0,95	0,96	0,97	0,97
0,2	0,97	0,96	0,95	0,93	0,92	0,9	0,89	0,88	0,88	0,88	0,89	0,9	0,92	0,93	0,95	0,96	0,97	0,97
0,19	0,97	0,96	0,95	0,94	0,92	0,9	0,89	0,89	0,88	0,89	0,89	0,9	0,92	0,94	0,95	0,96	0,97	0,98
0,18	0,98	0,97	0,95	0,94	0,92	0,91	0,89	0,89	0,89	0,89	0,89	0,9	0,92	0,94	0,95	0,97	0,98	0,98
0,17	0,98	0,97	0,96	0,94	0,92	0,91	0,9	0,89	0,89	0,89	0,9	0,91	0,92	0,94	0,96	0,97	0,98	0,98
0,16	0,98	0,97	0,96	0,94	0,92	0,91	0,9	0,89	0,89	0,89	0,9	0,91	0,92	0,94	0,96	0,97	0,98	0,98
0,15	0,98	0,97	0,96	0,94	0,93	0,91	0,9	0,89	0,89	0,89	0,9	0,91	0,93	0,94	0,96	0,97	0,98	0,99
0,14	0,98	0,97	0,96	0,94	0,93	0,91	0,9	0,89	0,89	0,89	0,9	0,91	0,93	0,94	0,96	0,97	0,98	0,99
0,13	0,99	0,98	0,96	0,95	0,93	0,91	0,9	0,89	0,89	0,89	0,9	0,91	0,93	0,95	0,96	0,98	0,99	0,99
0,12	0,99	0,98	0,96	0,95	0,93	0,92	0,9	0,9	0,89	0,9	0,9	0,92	0,93	0,95	0,96	0,98	0,99	0,99
0,11	0,99	0,98	0,97	0,95	0,93	0,92	0,9	0,9	0,89	0,9	0,9	0,92	0,93	0,95	0,97	0,98	0,99	0,99
0,1	0,99	0,98	0,97	0,95	0,93	0,92	0,9	0,9	0,9	0,9	0,91	0,92	0,93	0,95	0,97	0,98	0,99	0,99
0,09	0,99	0,98	0,97	0,95	0,93	0,92	0,91	0,9	0,9	0,9	0,91	0,92	0,93	0,95	0,97	0,98	0,99	0,99
0,08	0,99	0,98	0,97	0,95	0,93	0,92	0,91	0,9	0,9	0,9	0,91	0,92	0,93	0,95	0,97	0,98	0,99	1
0,07	0,99	0,98	0,97	0,95	0,94	0,92	0,91	0,9	0,9	0,9	0,91	0,92	0,94	0,95	0,97	0,98	0,99	1
0,06	0,99	0,98	0,97	0,95	0,94	0,92	0,91	0,9	0,9	0,9	0,91	0,92	0,94	0,95	0,97	0,98	0,99	1
0,05	0,99	0,99	0,97	0,95	0,94	0,92	0,91	0,9	0,9	0,9	0,91	0,92	0,94	0,95	0,97	0,99	0,99	1
0,04	1	0,99	0,97	0,95	0,94	0,92	0,91	0,9	0,9	0,9	0,91	0,92	0,94	0,95	0,97	0,99	1	1
0,03	1	0,99	0,97	0,95	0,94	0,92	0,91	0,9	0,9	0,9	0,91	0,92	0,94	0,95	0,97	0,99	1	1
0,02	1	0,99	0,97	0,95	0,94	0,92	0,91	0,9	0,9	0,9	0,91	0,92	0,94	0,95	0,97	0,99	1	1
0,01	1	0,99	0,97	0,95	0,94	0,92	0,91	0,9	0,9	0,9	0,91	0,92	0,94	0,95	0,97	0,99	1	1
0	1	0,99	0,97	0,95	0,94	0,92	0,91	0,9	0,9	0,9	0,91	0,92	0,94	0,95	0,97	0,99	1	1
-0,01	1	0,99	0,97	0,95	0,94	0,92	0,91	0,9	0,9	0,9	0,91	0,92	0,94	0,95	0,97	0,99	1	1
-0,02	1	0,99	0,97	0,95	0,94	0,92	0,91	0,9	0,9	0,9	0,91	0,92	0,94	0,95	0,97	0,99	1	1
-0,03	1	0,99	0,97	0,95	0,94	0,92	0,91	0,9	0,9	0,9	0,91	0,92	0,94	0,95	0,97	0,99	1	1
-0,04	1	0,99	0,97	0,95	0,94	0,92	0,91	0,9	0,9	0,9	0,91	0,92	0,94	0,95	0,97	0,99	1	1
-0,05	0,99	0,99	0,97	0,95	0,94	0,92	0,91	0,9	0,9	0,9	0,91	0,92	0,94	0,95	0,97	0,99	0,99	1
-0,06	0,99	0,98	0,97	0,95	0,94	0,92	0,91	0,9	0,9	0,9	0,91	0,92	0,94	0,95	0,97	0,98	0,99	1
-0,07	0,99	0,98	0,97	0,95	0,94	0,92	0,91	0,9	0,9	0,9	0,91	0,92	0,94	0,95	0,97	0,98	0,99	1
-0,08	0,99	0,98	0,97	0,95	0,93	0,92	0,91	0,9	0,9	0,9	0,91	0,92	0,93	0,95	0,97	0,98	0,99	1
-0,09	0,99	0,98	0,97	0,95	0,93	0,92	0,91	0,9	0,9	0,9	0,91	0,92	0,93	0,95	0,97	0,98	0,99	0,99
-0,1	0,99	0,98	0,97	0,95	0,93	0,92	0,91	0,9	0,9	0,9	0,91	0,92	0,93	0,95	0,97	0,98	0,99	0,99
-0,11	0,99	0,98	0,97	0,95	0,93	0,92	0,9	0,9	0,89	0,9	0,9	0,92	0,93	0,95	0,97	0,98	0,99	0,99
-0,12	0,99	0,98	0,96	0,95	0,93	0,92	0,9	0,9	0,89	0,9	0,9	0,92	0,93	0,95	0,96	0,98	0,99	0,99
-0,13	0,99	0,98	0,96	0,95	0,93	0,91	0,9	0,89	0,89	0,89	0,9	0,91	0,93	0,95	0,96	0,98	0,99	0,99
-0,14	0,98	0,97	0,96	0,94	0,93	0,91	0,9	0,89	0,89	0,89	0,9	0,91	0,93	0,94	0,96	0,97	0,98	0,99
-0,15	0,98	0,97	0,96	0,94	0,93	0,91	0,9	0,89	0,89	0,89	0,9	0,91	0,93	0,94	0,96	0,97	0,98	0,99
-0,16	0,98	0,97	0,96	0,94	0,92	0,91	0,9	0,89	0,89	0,89	0,9	0,91	0,92	0,94	0,96	0,97	0,98	0,98
-0,17	0,98	0,97	0,96	0,94	0,92	0,91	0,9	0,89	0,89	0,89	0,9	0,91	0,92	0,94	0,96	0,97	0,98	0,98
-0,18	0,98	0,97	0,95	0,94	0,92	0,91	0,89	0,89	0,89	0,89	0,89	0,91	0,92	0,94	0,95	0,97	0,98	0,98
-0,19	0,97	0,96	0,95	0,94	0,92	0,9	0,89	0,89	0,88	0,89	0,89	0,9	0,92	0,94	0,95	0,96	0,97	0,98
-0,2	0,97	0,96	0,95	0,93	0,92	0,9	0,89	0,88	0,88	0,88	0,89	0,9	0,92	0,93	0,95	0,96	0,97	0,97
-0,21	0,97	0,96	0,95	0,93	0,91	0,9	0,89	0,88	0,88	0,88	0,89	0,9	0,91	0,93	0,95	0,96	0,97	0,97
-0,22	0,97	0,96	0,94	0,93	0,91	0,9	0,89	0,88	0,88	0,88	0,89	0,9	0,91	0,93	0,94	0,95	0,97	0,97
-0,23	0,96	0,95	0,94	0,93	0,91	0,9	0,88	0,88	0,88	0,88	0,89	0,9	0,91	0,93	0,94	0,95	0,96	0,97
-0,24	0,96	0,95	0,94	0,92	0,91	0,89	0,88	0,88	0,87	0,88	0,88	0,89	0,91	0,92	0,94	0,95	0,96	0,96
-0,25	0,96	0,95	0,94	0,92	0,91	0,89	0,88	0,87	0,87	0,87	0,88	0,89	0,91	0,92	0,94	0,95	0,96	0,96
-0,26	0,95	0,95	0,93	0,92	0,9	0,89	0,88	0,87	0,87	0,87	0,88	0,89	0,9	0,92	0,93	0,95	0,95	0,96
-0,27	0,95	0,94	0,93	0,91	0,9	0,89	0,88	0,87	0,87	0,87	0,88	0,89	0,9	0,91	0,93	0,94	0,95	0,95
-0,28	0,95	0,94	0,93	0,91	0,9	0,88	0,87	0,87	0,86	0,87	0,87	0,88	0,9	0,91	0,93	0,94	0,95	0,95
-0,29	0,94	0,94	0,92	0,91	0,89	0,88	0,87	0,86	0,86	0,86	0,87	0,88	0,89	0,91	0,92	0,94	0,94	0,95
-0,3	0,94	0,93	0,92	0,91	0,89	0,88	0,87	0,86	0,86	0,86	0,87	0,88	0,89	0,91	0,92	0,93	0,94	0,94
-0,31	0,94	0,93	0,92	0,9	0,89	0,87	0,86	0,86	0,86	0,86	0,86	0,87	0,89	0,9	0,92	0,93	0,94	0,94
-0,32	0,93	0,92	0,91	0,9	0,88	0,87	0,86	0,85	0,85	0,85	0,86	0,87	0,88	0,9	0,91	0,92	0,93	0,93
-0,33	0,93	0,92	0,91	0,89	0,88	0,87	0,86	0,85	0,85	0,85	0,86	0,87	0,88	0,89	0,91	0,92	0,93	0,93
-0,34	0,92	0,92	0,9	0,89	0,88	0,86	0,85	0,85	0,85	0,85	0,85	0,86	0,88	0,89	0,9	0,92	0,92	0,93
-0,35	0,92	0,91	0,9	0,89	0,87	0,86	0,85	0,85	0,84	0,85	0,85	0,86	0,87	0,89	0,9	0,91	0,92	0,92
-0,36	0,91	0,91	0,9	0,88	0,87	0,86	0,85	0,84	0,84	0,84	0,85	0,86	0,87	0,88	0,9	0,91	0,91	0,92
-0,37	0,91	0,9	0,89	0,88	0,87	0,85	0,84	0,84	0,84	0,84	0,84	0,85	0,87	0,88	0,89	0,9	0,91	0,91
-0,38	0,9	0,9	0,89	0,87	0,86	0,85	0,84	0,83	0,83	0,83	0,84	0,85	0,86	0,87	0,89	0,9	0,9	0,91
-0,39	0,9	0,89	0,88	0,87	0,86	0,85	0,84	0,83	0,83	0,83	0,84	0,85	0,86	0,87	0,88	0,89	0,89	0,9
-0,4	0,89	0,89	0,88	0,86	0,85	0,84	0,83	0,83	0,82	0,83	0,83	0,84	0,85	0,86	0,88	0,89	0,89	0,9
-0,41	0,89	0,88	0,87	0,86	0,85	0,84	0,83	0,82	0,82	0,82	0,83	0,84	0,85	0,86	0,87	0,88	0,89	0,89
-0,42	0,88	0,88	0,87	0,85	0,84	0,83	0,82	0,82	0,82	0,82	0,82	0,83	0,84	0,85	0,87	0,88	0,88	0,88

-0,43	0,88	0,87	0,86	0,85	0,84	0,83	0,82	0,81	0,81	0,81	0,82	0,83	0,84	0,85	0,86	0,87	0,88	0,88
-0,44	0,87	0,86	0,86	0,84	0,83	0,82	0,82	0,81	0,81	0,81	0,82	0,82	0,83	0,84	0,86	0,86	0,87	0,87
-0,45	0,86	0,86	0,85	0,84	0,83	0,82	0,81	0,81	0,8	0,81	0,81	0,82	0,83	0,84	0,85	0,86	0,86	0,87
-0,46	0,86	0,85	0,84	0,83	0,82	0,81	0,81	0,8	0,8	0,81	0,81	0,82	0,83	0,84	0,85	0,86	0,86	0,86
-0,47	0,85	0,85	0,84	0,83	0,82	0,81	0,8	0,8	0,79	0,8	0,8	0,81	0,82	0,83	0,84	0,85	0,85	0,85
-0,48	0,84	0,84	0,83	0,82	0,81	0,8	0,8	0,79	0,79	0,79	0,8	0,8	0,81	0,82	0,83	0,84	0,84	0,85
-0,49	0,84	0,83	0,82	0,82	0,81	0,8	0,79	0,79	0,78	0,79	0,79	0,8	0,81	0,82	0,82	0,83	0,84	0,84
-0,5	0,83	0,83	0,82	0,81	0,8	0,79	0,79	0,78	0,78	0,78	0,79	0,79	0,8	0,81	0,82	0,83	0,83	0,83
-0,51	0,82	0,82	0,81	0,8	0,79	0,79	0,78	0,78	0,77	0,78	0,78	0,79	0,79	0,8	0,81	0,82	0,82	0,82
-0,52	0,81	0,81	0,8	0,8	0,79	0,78	0,77	0,77	0,77	0,77	0,77	0,78	0,79	0,8	0,8	0,81	0,81	0,82
-0,53	0,81	0,8	0,8	0,79	0,78	0,77	0,77	0,76	0,76	0,76	0,77	0,77	0,78	0,79	0,8	0,8	0,81	0,81
-0,54	0,8	0,8	0,79	0,78	0,78	0,77	0,76	0,76	0,76	0,76	0,76	0,77	0,78	0,78	0,79	0,8	0,8	0,8
-0,55	0,79	0,79	0,78	0,78	0,77	0,76	0,76	0,75	0,75	0,75	0,76	0,76	0,77	0,78	0,78	0,79	0,79	0,79
-0,56	0,78	0,78	0,77	0,77	0,76	0,76	0,75	0,75	0,75	0,75	0,75	0,76	0,76	0,77	0,77	0,78	0,78	0,78
-0,57	0,77	0,77	0,77	0,76	0,75	0,75	0,74	0,74	0,74	0,74	0,74	0,75	0,75	0,76	0,77	0,77	0,77	0,77
-0,58	0,76	0,76	0,76	0,75	0,75	0,74	0,74	0,73	0,73	0,73	0,74	0,74	0,75	0,75	0,76	0,76	0,76	0,76
-0,59	0,75	0,75	0,75	0,74	0,74	0,73	0,73	0,73	0,73	0,73	0,73	0,73	0,74	0,74	0,75	0,75	0,75	0,76
-0,6	0,74	0,74	0,74	0,74	0,73	0,73	0,72	0,72	0,72	0,72	0,72	0,73	0,73	0,74	0,74	0,74	0,74	0,75
-0,61	0,73	0,73	0,73	0,73	0,72	0,72	0,72	0,71	0,71	0,71	0,72	0,72	0,72	0,73	0,73	0,73	0,73	0,74
-0,62	0,72	0,72	0,72	0,72	0,72	0,71	0,71	0,71	0,71	0,71	0,71	0,71	0,72	0,72	0,72	0,72	0,72	0,72
-0,63	0,71	0,71	0,71	0,71	0,71	0,7	0,7	0,7	0,7	0,7	0,7	0,7	0,71	0,71	0,71	0,71	0,71	0,71
-0,64	0,7	0,7	0,7	0,7	0,7	0,7	0,69	0,69	0,69	0,69	0,69	0,7	0,7	0,7	0,7	0,7	0,7	0,7
-0,65	0,69	0,69	0,69	0,69	0,69	0,69	0,69	0,68	0,68	0,68	0,69	0,69	0,69	0,69	0,69	0,69	0,69	0,69
-0,66	0,68	0,68	0,68	0,68	0,68	0,68	0,68	0,68	0,68	0,68	0,68	0,68	0,68	0,68	0,68	0,68	0,68	0,68
-0,67	0,67	0,67	0,67	0,67	0,67	0,67	0,67	0,67	0,67	0,67	0,67	0,67	0,67	0,67	0,67	0,67	0,67	0,67
-0,68	0,66	0,66	0,66	0,66	0,66	0,66	0,66	0,66	0,66	0,66	0,66	0,66	0,66	0,66	0,66	0,66	0,66	0,66
-0,69	0,64	0,65	0,65	0,65	0,65	0,65	0,65	0,65	0,65	0,65	0,65	0,65	0,65	0,65	0,65	0,65	0,64	0,64
-0,7	0,63	0,63	0,64	0,64	0,64	0,64	0,64	0,64	0,64	0,64	0,64	0,64	0,64	0,64	0,64	0,63	0,63	0,63
-0,71	0,62	0,62	0,62	0,63	0,63	0,63	0,63	0,63	0,63	0,63	0,63	0,63	0,63	0,63	0,63	0,62	0,62	0,61
-0,72	0,6	0,61	0,61	0,62	0,62	0,62	0,62	0,62	0,62	0,62	0,62	0,62	0,62	0,62	0,62	0,61	0,61	0,6
-0,73	0,59	0,59	0,6	0,6	0,61	0,61	0,61	0,61	0,62	0,61	0,61	0,61	0,61	0,6	0,6	0,59	0,59	0,58
-0,74	0,57	0,58	0,58	0,59	0,6	0,6	0,6	0,61	0,6	0,6	0,6	0,6	0,6	0,59	0,58	0,58	0,57	0,57
-0,75	0,56	0,56	0,57	0,58	0,59	0,59	0,59	0,59	0,6	0,59	0,59	0,59	0,59	0,58	0,57	0,56	0,56	0,55
-0,76	0,54	0,55	0,56	0,56	0,57	0,58	0,58	0,58	0,58	0,58	0,58	0,58	0,57	0,56	0,56	0,55	0,54	0,54
-0,77	0,52	0,53	0,54	0,55	0,56	0,57	0,57	0,57	0,57	0,57	0,57	0,57	0,56	0,55	0,55	0,54	0,53	0,52
-0,78	0,5	0,51	0,52	0,54	0,55	0,55	0,56	0,56	0,56	0,56	0,56	0,55	0,55	0,54	0,52	0,51	0,5	0,5
-0,79	0,48	0,49	0,51	0,52	0,53	0,54	0,55	0,55	0,55	0,55	0,55	0,54	0,53	0,52	0,51	0,49	0,48	0,48
-0,8	0,46	0,47	0,49	0,5	0,52	0,53	0,53	0,54	0,54	0,54	0,53	0,53	0,52	0,5	0,49	0,47	0,46	0,46
-0,81	0,44	0,45	0,47	0,49	0,5	0,51	0,52	0,53	0,53	0,53	0,52	0,51	0,5	0,49	0,47	0,45	0,44	0,44
-0,82	0,42	0,43	0,45	0,47	0,49	0,5	0,51	0,52	0,52	0,52	0,51	0,5	0,49	0,47	0,45	0,43	0,42	0,41
-0,83	0,39	0,41	0,43	0,45	0,47	0,49	0,49	0,5	0,5	0,5	0,49	0,49	0,47	0,45	0,43	0,41	0,39	0,39
-0,84	0,37	0,38	0,41	0,43	0,45	0,47	0,48	0,49	0,49	0,49	0,48	0,47	0,45	0,43	0,41	0,38	0,37	0,36
-0,85	0,34	0,36	0,39	0,41	0,44	0,45	0,47	0,47	0,47	0,47	0,47	0,45	0,44	0,41	0,39	0,36	0,34	0,33
-0,86	0,3	0,33	0,36	0,39	0,42	0,44	0,45	0,46	0,46	0,46	0,45	0,44	0,42	0,39	0,36	0,33	0,3	0,29
-0,87	0,27	0,3	0,33	0,37	0,4	0,42	0,43	0,44	0,44	0,44	0,43	0,42	0,4	0,37	0,33	0,3	0,27	0,26
-0,88	0,22	0,26	0,3	0,34	0,37	0,4	0,42	0,43	0,42	0,42	0,42	0,4	0,37	0,34	0,3	0,26	0,22	0,21
-0,89	0,17	0,22	0,27	0,31	0,35	0,38	0,4	0,41	0,41	0,41	0,4	0,38	0,35	0,31	0,27	0,22	0,17	0,15
-0,9	0,08	0,16	0,23	0,28	0,33	0,36	0,38	0,39	0,39	0,39	0,38	0,36	0,33	0,28	0,23	0,16	0,08	
-0,91		0,07	0,18	0,25	0,3	0,33	0,36	0,37	0,37	0,37	0,36	0,33	0,3	0,25	0,18	0,07		
-0,92			0,11	0,21	0,27	0,31	0,33	0,35	0,35	0,35	0,33	0,31	0,27	0,21	0,11			
-0,93				0,16	0,23	0,28	0,31	0,33	0,33	0,33	0,31	0,28	0,23	0,16				
-0,94				0,07	0,19	0,25	0,28	0,3	0,31	0,3	0,28	0,25	0,19	0,07				
-0,95					0,13	0,21	0,25	0,27	0,28	0,27	0,25	0,21	0,13					
-0,96						0,17	0,22	0,24	0,25	0,24	0,22	0,17						
-0,97						0,1	0,18	0,21	0,22	0,21	0,18	0,1						
-0,98							0,12	0,17	0,18	0,17	0,12							
-0,99								0,11	0,13	0,11								
-1																		

$\sum_{\Delta\vec{h}_x} H_{\beta i/x}$	141	141	141	141	141	141	141	141	141	141	141	141	141	141	141	141	141	141
$n_{\beta,\Delta\vec{h}_x}=$	181	183	185	189	191	195	197	199	199	199	197	195	191	189	185	183	181	179
$\bar{h}_{\beta/x}=$	0,78	0,77	0,76	0,75	0,74	0,73	0,72	0,71	0,71	0,71	0,72	0,73	0,74	0,75	0,76	0,77	0,78	0,79
$d_{\beta/x}=$	1	0,99	0,97	0,95	0,94	0,92	0,91	0,9	0,9	0,9	0,91	0,92	0,94	0,95	0,97	0,99	1	1
$\bar{h}_\beta/d_\beta=$	0,78	0,78	0,79	0,78	0,79	0,79	0,79	0,79	0,79	0,79	0,79	0,79	0,79	0,78	0,79	0,78	0,78	0,79

$$\bar{\bar{h}}_{k\beta/x(\Delta\vec{h}_x)} = 0{,}75; \quad \widetilde{\bar{h}}_{k/x,\Delta\vec{h}_x} = 0{,}74; \quad \bar{\bar{h}}_{k\beta/x(\Delta\vec{h}_x)}/\bar{d}_{k/x} = 0{,}786 \to \pi/4 = 0{,}785398$$

$$k = 0{,}7$$

k 0,7 Δh/a	0,17	0,35	0,52	0,7	0,87	1,05	1,22	1,4	1,57	1,75	1,92	2,09	2,27	2,44	2,62	2,79	2,97	3,14
1																		
0,99								0,05	0,1	0,05								
0,98								0,11	0,14	0,11								
0,97								0,15	0,17	0,15								
0,96							0,1	0,18	0,2	0,18	0,1							
0,95							0,14	0,2	0,22	0,2	0,14							
0,94							0,18	0,23	0,24	0,23	0,18							
0,93						0,07	0,2	0,25	0,26	0,25	0,2	0,07						
0,92						0,13	0,23	0,26	0,27	0,26	0,23	0,13						
0,91						0,17	0,25	0,28	0,29	0,28	0,25	0,17						
0,9						0,2	0,27	0,3	0,31	0,3	0,27	0,2						
0,89						0,23	0,29	0,31	0,32	0,31	0,29	0,23						
0,88					0,11	0,25	0,3	0,33	0,33	0,33	0,3	0,25	0,11					
0,87					0,16	0,27	0,32	0,34	0,35	0,34	0,32	0,27	0,16					
0,86					0,2	0,29	0,33	0,35	0,36	0,35	0,33	0,29	0,2					
0,85					0,23	0,31	0,35	0,36	0,37	0,36	0,35	0,31	0,23					
0,84					0,26	0,33	0,36	0,38	0,38	0,38	0,36	0,33	0,26					
0,83				0,11	0,28	0,34	0,37	0,39	0,39	0,39	0,37	0,34	0,28	0,11				
0,82				0,17	0,3	0,36	0,39	0,4	0,4	0,4	0,39	0,36	0,3	0,17				
0,81				0,21	0,32	0,37	0,4	0,41	0,41	0,41	0,4	0,37	0,32	0,21				
0,8				0,25	0,34	0,39	0,41	0,42	0,42	0,42	0,41	0,39	0,34	0,25				
0,79				0,28	0,36	0,4	0,42	0,43	0,43	0,43	0,42	0,4	0,36	0,28				
0,78			0,11	0,3	0,38	0,41	0,43	0,44	0,44	0,44	0,43	0,41	0,38	0,3	0,11			
0,77			0,18	0,33	0,39	0,42	0,44	0,44	0,45	0,44	0,44	0,42	0,39	0,33	0,18			
0,76			0,23	0,35	0,41	0,44	0,45	0,45	0,45	0,45	0,44	0,44	0,41	0,35	0,23			
0,75			0,27	0,37	0,42	0,45	0,46	0,46	0,46	0,46	0,46	0,45	0,42	0,37	0,27			
0,74		0,06	0,3	0,39	0,44	0,46	0,47	0,47	0,47	0,47	0,47	0,46	0,44	0,39	0,3	0,06		
0,73		0,16	0,33	0,41	0,45	0,47	0,48	0,48	0,48	0,48	0,48	0,47	0,45	0,41	0,33	0,16		
0,72		0,23	0,36	0,43	0,46	0,48	0,48	0,49	0,49	0,49	0,48	0,48	0,46	0,43	0,36	0,23		
0,71	0,05	0,27	0,38	0,44	0,47	0,49	0,49	0,49	0,49	0,49	0,49	0,49	0,47	0,44	0,38	0,27	0,05	
0,7	0,17	0,31	0,4	0,46	0,49	0,5	0,5	0,5	0,5	0,5	0,5	0,5	0,49	0,46	0,4	0,31	0,17	
0,69	0,24	0,35	0,43	0,47	0,5	0,51	0,51	0,51	0,51	0,51	0,51	0,51	0,5	0,47	0,43	0,35	0,24	0,17
0,68	0,29	0,38	0,45	0,49	0,51	0,51	0,51	0,51	0,51	0,51	0,51	0,51	0,51	0,49	0,45	0,38	0,29	0,24
0,67	0,33	0,4	0,47	0,5	0,52	0,52	0,52	0,52	0,52	0,52	0,52	0,52	0,52	0,5	0,47	0,4	0,33	0,29
0,66	0,37	0,43	0,48	0,51	0,53	0,53	0,53	0,53	0,53	0,53	0,53	0,53	0,53	0,51	0,48	0,43	0,37	0,33
0,65	0,4	0,45	0,5	0,53	0,54	0,54	0,54	0,53	0,53	0,53	0,54	0,54	0,54	0,53	0,5	0,45	0,4	0,37
0,64	0,43	0,48	0,52	0,54	0,55	0,55	0,54	0,54	0,54	0,54	0,54	0,55	0,55	0,54	0,52	0,48	0,43	0,41
0,63	0,46	0,5	0,53	0,55	0,56	0,55	0,55	0,55	0,54	0,55	0,55	0,55	0,56	0,55	0,53	0,5	0,46	0,44
0,62	0,48	0,52	0,55	0,56	0,56	0,56	0,56	0,55	0,55	0,55	0,56	0,56	0,56	0,56	0,55	0,52	0,48	0,46
0,61	0,51	0,54	0,56	0,57	0,57	0,57	0,56	0,56	0,55	0,56	0,56	0,57	0,57	0,57	0,56	0,54	0,51	0,49
0,6	0,53	0,55	0,58	0,58	0,58	0,57	0,57	0,56	0,56	0,56	0,57	0,57	0,58	0,58	0,58	0,55	0,53	0,52
0,59	0,55	0,57	0,59	0,59	0,59	0,58	0,57	0,57	0,57	0,57	0,57	0,58	0,59	0,59	0,59	0,57	0,55	0,54
0,58	0,57	0,59	0,6	0,6	0,6	0,59	0,58	0,57	0,57	0,57	0,58	0,59	0,6	0,6	0,6	0,59	0,57	0,56
0,57	0,59	0,6	0,61	0,61	0,6	0,59	0,58	0,58	0,58	0,58	0,58	0,59	0,6	0,61	0,61	0,6	0,59	0,58
0,56	0,61	0,62	0,62	0,62	0,61	0,6	0,59	0,58	0,58	0,58	0,59	0,6	0,61	0,62	0,62	0,62	0,61	0,6
0,55	0,62	0,63	0,64	0,63	0,62	0,61	0,59	0,59	0,58	0,59	0,59	0,61	0,62	0,63	0,64	0,63	0,62	0,61
0,54	0,64	0,65	0,65	0,64	0,63	0,61	0,6	0,59	0,59	0,59	0,6	0,61	0,63	0,64	0,65	0,65	0,64	0,64
0,53	0,66	0,66	0,66	0,65	0,63	0,62	0,6	0,6	0,6	0,6	0,6	0,62	0,63	0,65	0,66	0,66	0,66	0,65
0,52	0,67	0,67	0,67	0,66	0,64	0,62	0,61	0,6	0,6	0,6	0,61	0,62	0,64	0,66	0,67	0,67	0,67	0,67
0,51	0,69	0,69	0,68	0,66	0,65	0,63	0,61	0,61	0,6	0,61	0,61	0,63	0,65	0,66	0,68	0,69	0,69	0,68
0,5	0,7	0,7	0,69	0,67	0,65	0,63	0,62	0,61	0,61	0,61	0,62	0,63	0,65	0,67	0,69	0,7	0,7	0,7
0,49	0,71	0,71	0,7	0,68	0,66	0,64	0,62	0,62	0,61	0,61	0,62	0,64	0,66	0,68	0,7	0,71	0,71	0,71
0,48	0,73	0,72	0,71	0,69	0,66	0,64	0,63	0,62	0,61	0,62	0,63	0,64	0,66	0,69	0,71	0,72	0,73	0,73
0,47	0,74	0,73	0,71	0,69	0,67	0,65	0,63	0,62	0,62	0,62	0,63	0,65	0,67	0,69	0,71	0,73	0,74	0,74
0,46	0,75	0,74	0,72	0,7	0,67	0,65	0,64	0,63	0,62	0,63	0,64	0,65	0,67	0,7	0,72	0,74	0,75	0,75
0,45	0,76	0,75	0,73	0,71	0,68	0,66	0,64	0,63	0,63	0,63	0,64	0,66	0,68	0,71	0,73	0,75	0,76	0,77
0,44	0,77	0,76	0,74	0,71	0,68	0,66	0,64	0,63	0,63	0,63	0,64	0,66	0,68	0,71	0,74	0,76	0,77	0,78
0,43	0,78	0,77	0,75	0,72	0,69	0,67	0,65	0,64	0,63	0,64	0,65	0,67	0,69	0,72	0,75	0,77	0,78	0,79
0,42	0,79	0,78	0,75	0,72	0,69	0,67	0,65	0,64	0,64	0,64	0,65	0,67	0,69	0,72	0,75	0,78	0,79	0,8
0,41	0,8	0,79	0,76	0,73	0,7	0,67	0,65	0,64	0,64	0,64	0,65	0,67	0,7	0,73	0,76	0,79	0,8	0,81
0,4	0,81	0,79	0,77	0,73	0,7	0,68	0,66	0,65	0,64	0,65	0,66	0,68	0,7	0,73	0,77	0,79	0,81	0,82
0,39	0,82	0,8	0,77	0,74	0,71	0,68	0,66	0,65	0,65	0,65	0,66	0,68	0,71	0,74	0,77	0,8	0,82	0,83
0,38	0,83	0,81	0,78	0,75	0,71	0,68	0,66	0,65	0,65	0,65	0,66	0,68	0,71	0,75	0,78	0,81	0,83	0,84
0,37	0,84	0,82	0,79	0,75	0,72	0,69	0,67	0,65	0,65	0,65	0,67	0,69	0,72	0,75	0,79	0,82	0,84	0,85
0,36	0,85	0,83	0,79	0,75	0,72	0,69	0,67	0,66	0,65	0,66	0,67	0,69	0,72	0,75	0,79	0,83	0,85	0,86
0,35	0,86	0,83	0,8	0,76	0,72	0,69	0,67	0,66	0,66	0,66	0,67	0,69	0,72	0,76	0,8	0,83	0,86	0,87
0,34	0,86	0,84	0,8	0,76	0,73	0,7	0,68	0,66	0,66	0,66	0,68	0,7	0,73	0,76	0,8	0,84	0,86	0,87

0,33	0,87	0,85	0,81	0,77	0,73	0,7	0,68	0,67	0,66	0,67	0,68	0,7	0,73	0,77	0,81	0,85	0,87	0,88
0,32	0,88	0,85	0,81	0,77	0,74	0,7	0,68	0,67	0,66	0,67	0,68	0,7	0,74	0,77	0,81	0,85	0,88	0,89
0,31	0,89	0,86	0,82	0,78	0,74	0,71	0,68	0,67	0,67	0,67	0,68	0,71	0,74	0,78	0,82	0,86	0,89	0,9
0,3	0,89	0,86	0,82	0,78	0,74	0,71	0,69	0,67	0,67	0,67	0,69	0,71	0,74	0,78	0,82	0,86	0,89	0,9
0,29	0,9	0,87	0,83	0,78	0,74	0,71	0,69	0,67	0,67	0,67	0,69	0,71	0,74	0,78	0,83	0,87	0,9	0,91
0,28	0,91	0,87	0,83	0,79	0,75	0,71	0,69	0,68	0,67	0,68	0,69	0,71	0,75	0,79	0,83	0,87	0,91	0,92
0,27	0,91	0,88	0,84	0,79	0,75	0,72	0,69	0,68	0,67	0,68	0,69	0,72	0,75	0,79	0,84	0,88	0,91	0,92
0,26	0,92	0,88	0,84	0,79	0,75	0,72	0,7	0,68	0,68	0,68	0,7	0,72	0,75	0,79	0,84	0,88	0,92	0,93
0,25	0,92	0,89	0,84	0,8	0,76	0,72	0,7	0,68	0,68	0,68	0,7	0,72	0,76	0,8	0,84	0,89	0,92	0,93
0,24	0,93	0,89	0,85	0,8	0,76	0,72	0,7	0,68	0,68	0,68	0,7	0,72	0,76	0,8	0,85	0,89	0,93	0,94
0,23	0,93	0,9	0,85	0,8	0,76	0,73	0,7	0,69	0,68	0,69	0,7	0,73	0,76	0,8	0,85	0,9	0,93	0,94
0,22	0,94	0,9	0,86	0,81	0,76	0,73	0,7	0,69	0,68	0,69	0,7	0,73	0,76	0,81	0,86	0,9	0,94	0,95
0,21	0,94	0,91	0,86	0,81	0,77	0,73	0,7	0,69	0,68	0,69	0,7	0,73	0,77	0,81	0,86	0,91	0,94	0,95
0,2	0,94	0,91	0,86	0,81	0,77	0,73	0,71	0,69	0,69	0,69	0,71	0,73	0,77	0,81	0,86	0,91	0,94	0,96
0,19	0,95	0,91	0,86	0,81	0,77	0,73	0,71	0,69	0,69	0,69	0,71	0,73	0,77	0,81	0,86	0,91	0,95	0,96
0,18	0,95	0,92	0,87	0,82	0,77	0,74	0,71	0,69	0,69	0,69	0,71	0,74	0,77	0,82	0,87	0,92	0,95	0,97
0,17	0,96	0,92	0,87	0,82	0,77	0,74	0,71	0,7	0,69	0,7	0,71	0,74	0,77	0,82	0,87	0,92	0,96	0,97
0,16	0,96	0,92	0,87	0,82	0,78	0,74	0,71	0,7	0,69	0,7	0,71	0,74	0,78	0,82	0,87	0,92	0,96	0,97
0,15	0,96	0,92	0,87	0,82	0,78	0,74	0,71	0,7	0,69	0,7	0,71	0,74	0,78	0,82	0,87	0,92	0,96	0,98
0,14	0,97	0,93	0,88	0,82	0,78	0,74	0,71	0,7	0,69	0,7	0,71	0,74	0,78	0,82	0,88	0,93	0,97	0,98
0,13	0,97	0,93	0,88	0,83	0,78	0,74	0,72	0,7	0,69	0,7	0,72	0,74	0,78	0,83	0,88	0,93	0,97	0,98
0,12	0,97	0,93	0,88	0,83	0,78	0,74	0,72	0,7	0,69	0,7	0,72	0,74	0,78	0,83	0,88	0,93	0,97	0,99
0,11	0,97	0,93	0,88	0,83	0,78	0,74	0,72	0,7	0,7	0,7	0,72	0,74	0,78	0,83	0,88	0,93	0,97	0,99
0,1	0,97	0,94	0,88	0,83	0,78	0,75	0,72	0,7	0,7	0,7	0,72	0,75	0,78	0,83	0,88	0,94	0,97	0,99
0,09	0,98	0,94	0,88	0,83	0,78	0,75	0,72	0,7	0,7	0,7	0,72	0,75	0,78	0,83	0,88	0,94	0,98	0,99
0,08	0,98	0,94	0,89	0,83	0,78	0,75	0,72	0,7	0,7	0,7	0,72	0,75	0,78	0,83	0,89	0,94	0,98	0,99
0,07	0,98	0,94	0,89	0,83	0,79	0,75	0,72	0,7	0,7	0,7	0,72	0,75	0,79	0,83	0,89	0,94	0,98	0,99
0,06	0,98	0,94	0,89	0,83	0,79	0,75	0,72	0,7	0,7	0,7	0,72	0,75	0,79	0,83	0,89	0,94	0,98	1
0,05	0,98	0,94	0,89	0,83	0,79	0,75	0,72	0,7	0,7	0,7	0,72	0,75	0,79	0,83	0,89	0,94	0,98	1
0,04	0,98	0,94	0,89	0,84	0,79	0,75	0,72	0,7	0,7	0,7	0,72	0,75	0,79	0,84	0,89	0,94	0,98	1
0,03	0,98	0,94	0,89	0,84	0,79	0,75	0,72	0,71	0,7	0,71	0,72	0,75	0,79	0,84	0,89	0,94	0,98	1
0,02	0,98	0,94	0,89	0,84	0,79	0,75	0,72	0,71	0,7	0,71	0,72	0,75	0,79	0,84	0,89	0,94	0,98	1
0,01	0,98	0,94	0,89	0,84	0,79	0,75	0,72	0,71	0,7	0,71	0,72	0,75	0,79	0,84	0,89	0,94	0,98	1
-0	0,98	0,94	0,89	0,84	0,79	0,75	0,72	0,71	0,7	0,71	0,72	0,75	0,79	0,84	0,89	0,94	0,98	1
-0,01	0,98	0,94	0,89	0,84	0,79	0,75	0,72	0,71	0,7	0,71	0,72	0,75	0,79	0,84	0,89	0,94	0,98	1
-0,02	0,98	0,94	0,89	0,84	0,79	0,75	0,72	0,71	0,7	0,71	0,72	0,75	0,79	0,84	0,89	0,94	0,98	1
-0,03	0,98	0,94	0,89	0,84	0,79	0,75	0,72	0,71	0,7	0,71	0,72	0,75	0,79	0,84	0,89	0,94	0,98	1
-0,04	0,98	0,94	0,89	0,84	0,79	0,75	0,72	0,7	0,7	0,7	0,72	0,75	0,79	0,84	0,89	0,94	0,98	1
-0,05	0,98	0,94	0,89	0,83	0,79	0,75	0,72	0,7	0,7	0,7	0,72	0,75	0,79	0,83	0,89	0,94	0,98	1
-0,06	0,98	0,94	0,89	0,83	0,79	0,75	0,72	0,7	0,7	0,7	0,72	0,75	0,79	0,83	0,89	0,94	0,98	1
-0,07	0,98	0,94	0,89	0,83	0,79	0,75	0,72	0,7	0,7	0,7	0,72	0,75	0,79	0,83	0,89	0,94	0,98	0,99
-0,08	0,98	0,94	0,89	0,83	0,78	0,75	0,72	0,7	0,7	0,7	0,72	0,75	0,78	0,83	0,89	0,94	0,98	0,99
-0,09	0,98	0,94	0,88	0,83	0,78	0,75	0,72	0,7	0,7	0,7	0,72	0,75	0,78	0,83	0,88	0,94	0,98	0,99
-0,1	0,97	0,94	0,88	0,83	0,78	0,75	0,72	0,7	0,7	0,7	0,72	0,75	0,78	0,83	0,88	0,94	0,97	0,99
-0,11	0,97	0,93	0,88	0,83	0,78	0,74	0,72	0,7	0,7	0,7	0,72	0,74	0,78	0,83	0,88	0,93	0,97	0,99
-0,12	0,97	0,93	0,88	0,83	0,78	0,74	0,72	0,7	0,69	0,7	0,72	0,74	0,78	0,83	0,88	0,93	0,97	0,99
-0,13	0,97	0,93	0,88	0,83	0,78	0,74	0,72	0,7	0,69	0,7	0,72	0,74	0,78	0,83	0,88	0,93	0,97	0,98
-0,14	0,97	0,93	0,88	0,82	0,78	0,74	0,71	0,7	0,69	0,7	0,71	0,74	0,78	0,82	0,88	0,93	0,97	0,98
-0,15	0,96	0,92	0,87	0,82	0,78	0,74	0,71	0,7	0,69	0,7	0,71	0,74	0,78	0,82	0,87	0,92	0,96	0,98
-0,16	0,96	0,92	0,87	0,82	0,78	0,74	0,71	0,7	0,69	0,7	0,71	0,74	0,78	0,82	0,87	0,92	0,96	0,97
-0,17	0,96	0,92	0,87	0,82	0,77	0,74	0,71	0,7	0,69	0,7	0,71	0,74	0,77	0,82	0,87	0,92	0,96	0,97
-0,18	0,95	0,92	0,87	0,82	0,77	0,74	0,71	0,69	0,69	0,69	0,71	0,74	0,77	0,82	0,87	0,92	0,95	0,97
-0,19	0,95	0,91	0,86	0,81	0,77	0,73	0,71	0,69	0,69	0,69	0,71	0,73	0,77	0,81	0,86	0,91	0,95	0,96
-0,2	0,94	0,91	0,86	0,81	0,77	0,73	0,71	0,69	0,69	0,69	0,71	0,73	0,77	0,81	0,86	0,91	0,94	0,96
-0,21	0,94	0,91	0,86	0,81	0,77	0,73	0,7	0,69	0,68	0,69	0,7	0,73	0,77	0,81	0,86	0,91	0,94	0,95
-0,22	0,94	0,9	0,86	0,81	0,76	0,73	0,7	0,69	0,68	0,69	0,7	0,73	0,76	0,81	0,86	0,9	0,94	0,95
-0,23	0,93	0,9	0,85	0,8	0,76	0,73	0,7	0,69	0,68	0,69	0,7	0,73	0,76	0,8	0,85	0,9	0,93	0,94
-0,24	0,93	0,89	0,85	0,8	0,76	0,72	0,7	0,68	0,68	0,68	0,7	0,72	0,76	0,8	0,85	0,89	0,93	0,94
-0,25	0,92	0,89	0,84	0,8	0,76	0,72	0,7	0,68	0,68	0,68	0,7	0,72	0,76	0,8	0,84	0,89	0,92	0,93
-0,26	0,92	0,88	0,84	0,79	0,75	0,72	0,7	0,68	0,68	0,68	0,7	0,72	0,75	0,79	0,84	0,88	0,92	0,93
-0,27	0,91	0,88	0,84	0,79	0,75	0,72	0,69	0,68	0,67	0,68	0,69	0,72	0,75	0,79	0,84	0,88	0,91	0,92
-0,28	0,91	0,87	0,83	0,79	0,75	0,71	0,69	0,68	0,67	0,68	0,69	0,71	0,75	0,79	0,83	0,87	0,91	0,92
-0,29	0,9	0,87	0,83	0,78	0,74	0,71	0,69	0,67	0,67	0,67	0,69	0,71	0,74	0,78	0,83	0,87	0,9	0,91
-0,3	0,89	0,86	0,82	0,78	0,74	0,71	0,69	0,67	0,67	0,67	0,69	0,71	0,74	0,78	0,82	0,86	0,89	0,9
-0,31	0,89	0,86	0,82	0,78	0,74	0,71	0,68	0,67	0,67	0,67	0,68	0,71	0,74	0,78	0,82	0,86	0,89	0,9
-0,32	0,88	0,85	0,81	0,77	0,74	0,7	0,68	0,67	0,66	0,67	0,68	0,7	0,74	0,77	0,81	0,85	0,88	0,89
-0,33	0,87	0,85	0,81	0,77	0,73	0,7	0,68	0,67	0,66	0,67	0,68	0,7	0,73	0,77	0,81	0,85	0,87	0,88
-0,34	0,86	0,84	0,8	0,76	0,73	0,7	0,68	0,66	0,66	0,66	0,68	0,7	0,73	0,76	0,8	0,84	0,86	0,87
-0,35	0,86	0,83	0,8	0,76	0,72	0,69	0,67	0,66	0,66	0,66	0,67	0,69	0,72	0,76	0,8	0,83	0,86	0,87
-0,36	0,85	0,83	0,79	0,75	0,72	0,69	0,67	0,66	0,65	0,66	0,67	0,69	0,72	0,75	0,79	0,83	0,85	0,86
-0,37	0,84	0,82	0,79	0,75	0,72	0,69	0,67	0,65	0,65	0,65	0,67	0,69	0,72	0,75	0,79	0,82	0,84	0,85
-0,38	0,83	0,81	0,78	0,75	0,71	0,68	0,66	0,65	0,65	0,65	0,66	0,68	0,71	0,75	0,78	0,81	0,83	0,84
-0,39	0,82	0,8	0,77	0,74	0,71	0,68	0,66	0,65	0,64	0,65	0,66	0,68	0,71	0,74	0,77	0,8	0,82	0,83
-0,4	0,81	0,79	0,77	0,73	0,7	0,68	0,66	0,65	0,64	0,65	0,66	0,68	0,7	0,73	0,77	0,79	0,81	0,82
-0,41	0,8	0,79	0,76	0,73	0,7	0,67	0,65	0,64	0,64	0,64	0,65	0,67	0,7	0,73	0,76	0,79	0,8	0,81
-0,42	0,79	0,78	0,75	0,72	0,69	0,67	0,65	0,64	0,64	0,64	0,65	0,67	0,69	0,72	0,75	0,78	0,79	0,8

	1	2	3	4	5	6	7	8	9	10	11	12	13	14	15	16	17	18
-0,43	0,78	0,77	0,75	0,72	0,69	0,67	0,65	0,64	0,63	0,64	0,65	0,67	0,69	0,72	0,75	0,77	0,78	0,79
-0,44	0,77	0,76	0,74	0,71	0,68	0,66	0,64	0,63	0,63	0,63	0,64	0,66	0,68	0,71	0,74	0,76	0,77	0,78
-0,45	0,76	0,75	0,73	0,71	0,68	0,66	0,64	0,63	0,63	0,63	0,64	0,66	0,68	0,71	0,73	0,75	0,76	0,77
-0,46	0,75	0,74	0,72	0,7	0,67	0,65	0,64	0,63	0,62	0,63	0,64	0,65	0,67	0,7	0,72	0,74	0,75	0,75
-0,47	0,74	0,73	0,71	0,69	0,67	0,65	0,63	0,62	0,62	0,62	0,63	0,65	0,67	0,69	0,71	0,73	0,74	0,74
-0,48	0,73	0,72	0,71	0,69	0,66	0,64	0,63	0,62	0,61	0,62	0,63	0,64	0,66	0,69	0,71	0,72	0,73	0,73
-0,49	0,71	0,71	0,7	0,68	0,66	0,64	0,62	0,61	0,61	0,61	0,62	0,64	0,66	0,68	0,7	0,71	0,71	0,71
-0,5	0,7	0,7	0,69	0,67	0,65	0,63	0,62	0,61	0,61	0,61	0,62	0,63	0,65	0,67	0,69	0,7	0,7	0,7
-0,51	0,69	0,69	0,68	0,66	0,65	0,63	0,61	0,61	0,6	0,61	0,61	0,63	0,65	0,66	0,68	0,69	0,69	0,68
-0,52	0,67	0,67	0,67	0,66	0,64	0,62	0,61	0,6	0,6	0,6	0,61	0,62	0,64	0,66	0,67	0,67	0,67	0,67
-0,53	0,66	0,66	0,66	0,65	0,63	0,62	0,6	0,6	0,59	0,6	0,6	0,62	0,63	0,65	0,66	0,66	0,66	0,65
-0,54	0,64	0,65	0,65	0,64	0,63	0,61	0,6	0,59	0,59	0,59	0,6	0,61	0,63	0,64	0,65	0,65	0,64	0,64
-0,55	0,62	0,63	0,64	0,63	0,62	0,61	0,59	0,59	0,58	0,59	0,59	0,61	0,62	0,63	0,64	0,63	0,62	0,62
-0,56	0,61	0,62	0,62	0,62	0,61	0,6	0,59	0,58	0,58	0,58	0,59	0,6	0,61	0,62	0,62	0,62	0,61	0,6
-0,57	0,59	0,6	0,61	0,61	0,6	0,59	0,58	0,58	0,58	0,58	0,58	0,59	0,6	0,61	0,61	0,6	0,59	0,58
-0,58	0,57	0,59	0,6	0,6	0,6	0,59	0,58	0,57	0,57	0,57	0,58	0,59	0,6	0,6	0,6	0,59	0,57	0,56
-0,59	0,55	0,57	0,59	0,59	0,59	0,58	0,57	0,57	0,57	0,57	0,57	0,58	0,59	0,59	0,59	0,57	0,55	0,54
-0,6	0,53	0,55	0,58	0,58	0,58	0,57	0,57	0,56	0,56	0,56	0,57	0,57	0,58	0,58	0,58	0,55	0,53	0,52
-0,61	0,51	0,54	0,56	0,57	0,57	0,57	0,56	0,56	0,55	0,56	0,56	0,57	0,57	0,57	0,56	0,54	0,51	0,49
-0,62	0,48	0,52	0,55	0,56	0,56	0,56	0,56	0,55	0,55	0,55	0,56	0,56	0,56	0,56	0,55	0,52	0,48	0,46
-0,63	0,46	0,5	0,53	0,55	0,56	0,55	0,55	0,55	0,54	0,55	0,55	0,55	0,56	0,55	0,53	0,5	0,46	0,44
-0,64	0,43	0,48	0,52	0,54	0,55	0,55	0,54	0,54	0,54	0,54	0,54	0,55	0,55	0,54	0,52	0,48	0,43	0,41
-0,65	0,4	0,45	0,5	0,53	0,54	0,54	0,54	0,53	0,53	0,53	0,54	0,54	0,54	0,53	0,5	0,45	0,4	0,37
-0,66	0,37	0,43	0,48	0,51	0,53	0,53	0,53	0,53	0,53	0,53	0,53	0,53	0,53	0,51	0,48	0,43	0,37	0,33
-0,67	0,33	0,4	0,47	0,5	0,52	0,52	0,52	0,52	0,52	0,52	0,52	0,52	0,52	0,5	0,47	0,4	0,33	0,29
-0,68	0,29	0,38	0,45	0,49	0,51	0,51	0,51	0,51	0,51	0,51	0,51	0,51	0,51	0,49	0,45	0,38	0,29	0,24
-0,69	0,24	0,35	0,43	0,47	0,5	0,51	0,51	0,51	0,51	0,51	0,51	0,51	0,5	0,47	0,43	0,35	0,24	0,17
-0,7	0,17	0,31	0,4	0,46	0,49	0,5	0,5	0,5	0,5	0,5	0,5	0,5	0,49	0,46	0,4	0,31	0,17	
-0,71	0,05	0,27	0,38	0,44	0,47	0,49	0,49	0,49	0,49	0,49	0,49	0,49	0,47	0,44	0,38	0,27	0,05	
-0,72		0,23	0,36	0,43	0,46	0,48	0,48	0,49	0,49	0,49	0,48	0,48	0,46	0,43	0,36	0,23		
-0,73		0,16	0,33	0,41	0,45	0,47	0,48	0,48	0,48	0,48	0,48	0,47	0,45	0,41	0,33	0,16		
-0,74		0,06	0,3	0,39	0,44	0,46	0,47	0,47	0,47	0,47	0,47	0,46	0,44	0,39	0,3	0,06		
-0,75			0,27	0,37	0,42	0,45	0,46	0,46	0,46	0,46	0,46	0,45	0,42	0,37	0,27			
-0,76			0,23	0,35	0,41	0,44	0,45	0,45	0,45	0,45	0,45	0,44	0,41	0,35	0,23			
-0,77			0,18	0,33	0,39	0,42	0,44	0,44	0,45	0,44	0,44	0,42	0,39	0,33	0,18			
-0,78			0,11	0,3	0,38	0,41	0,43	0,44	0,44	0,44	0,43	0,41	0,38	0,3	0,11			
-0,79				0,28	0,36	0,4	0,42	0,43	0,43	0,43	0,42	0,4	0,36	0,28				
-0,8				0,25	0,34	0,39	0,41	0,42	0,42	0,42	0,41	0,39	0,34	0,25				
-0,81				0,21	0,32	0,37	0,4	0,41	0,41	0,41	0,4	0,37	0,32	0,21				
-0,82				0,17	0,3	0,36	0,4	0,4	0,4	0,4	0,39	0,36	0,3	0,17				
-0,83				0,11	0,28	0,34	0,37	0,39	0,39	0,39	0,37	0,34	0,28	0,11				
-0,84					0,26	0,33	0,36	0,38	0,38	0,38	0,36	0,33	0,26					
-0,85					0,23	0,31	0,35	0,36	0,37	0,36	0,35	0,31	0,23					
-0,86					0,2	0,29	0,33	0,35	0,36	0,35	0,33	0,29	0,2					
-0,87					0,16	0,27	0,32	0,34	0,35	0,34	0,32	0,27	0,16					
-0,88					0,11	0,25	0,3	0,33	0,33	0,33	0,3	0,25	0,11					
-0,89						0,23	0,29	0,31	0,32	0,31	0,29	0,23						
-0,9						0,2	0,27	0,3	0,31	0,3	0,27	0,2						
-0,91						0,17	0,25	0,28	0,29	0,28	0,25	0,17						
-0,92						0,13	0,23	0,26	0,27	0,26	0,23	0,13						
-0,93						0,07	0,2	0,25	0,26	0,25	0,2	0,07						
-0,94							0,18	0,23	0,24	0,23	0,18							
-0,95							0,14	0,2	0,22	0,2	0,14							
-0,96							0,1	0,18	0,2	0,18	0,1							
-0,97								0,15	0,17	0,15								
-0,98								0,11	0,14	0,11								
-0,99								0,05	0,1	0,05								
-1																		

	1	2	3	4	5	6	7	8	9	10	11	12	13	14	15	16	17	18
$\sum_{\Delta\vec{h}_x} H_{\beta i/x}$	110	110	110	110	110	110	110	110	110	110	110	110	110	110	110	110	110	110
$n_{\beta,\Delta\vec{h}_x} =$	143	149	157	167	177	187	193	199	199	199	193	187	177	167	157	149	143	139
$\bar{h}_{\beta/x} =$	0,77	0,74	0,7	0,66	0,62	0,59	0,57	0,55	0,55	0,55	0,57	0,59	0,62	0,66	0,7	0,74	0,77	0,79
$d_{\beta/x} =$	0,98	0,94	0,89	0,84	0,79	0,75	0,72	0,71	0,7	0,71	0,72	0,75	0,79	0,84	0,89	0,94	0,98	1
$\bar{h}_\beta/d_\beta =$	0,78	0,78	0,79	0,79	0,79	0,78	0,79	0,78	0,79	0,78	0,79	0,79	0,79	0,79	0,78	0,78	0,79	

$$\bar{\bar{h}}_{k\beta/x(\Delta\vec{h}_x)} = 0{,}65; \quad \tilde{\bar{h}}_{k/x,\Delta\vec{h}_x} = 0{,}64; \quad \bar{\bar{h}}_{k\beta/x(\Delta\vec{h}_x)}/\bar{d}_{k/x} = 0{,}786 \rightarrow \pi/4 = 0{,}785398$$

$$k = 0,5$$

0,5 Δh/a \ β	0,17	0,35	0,52	0,7	0,87	1,05	1,22	1,4	1,57	1,75	1,92	2,09	2,27	2,44	2,62	2,79	2,97	3,14
1																		
0,99									0,07									
0,98								0,07	0,1	0,07								
0,97								0,1	0,12	0,1								
0,96								0,12	0,14	0,12								
0,95							0,05	0,14	0,16	0,14	0,05							
0,94							0,09	0,16	0,17	0,16	0,09							
0,93							0,12	0,17	0,18	0,17	0,12							
0,92							0,14	0,19	0,2	0,19	0,14							
0,91							0,16	0,2	0,21	0,2	0,16							
0,9						0,03	0,18	0,21	0,22	0,21	0,18	0,03						
0,89						0,09	0,19	0,22	0,23	0,22	0,19	0,09						
0,88						0,12	0,2	0,23	0,24	0,23	0,2	0,12						
0,87						0,15	0,22	0,24	0,25	0,24	0,22	0,15						
0,86						0,17	0,23	0,25	0,26	0,25	0,23	0,17						
0,85						0,18	0,24	0,26	0,26	0,26	0,24	0,18						
0,84						0,2	0,25	0,27	0,27	0,27	0,25	0,2						
0,83					0,03	0,22	0,26	0,27	0,28	0,27	0,26	0,22	0,03					
0,82					0,1	0,23	0,27	0,28	0,29	0,28	0,27	0,23	0,1					
0,81					0,13	0,24	0,28	0,29	0,29	0,29	0,28	0,24	0,13					
0,8					0,16	0,26	0,29	0,3	0,3	0,3	0,29	0,26	0,16					
0,79					0,19	0,27	0,29	0,3	0,31	0,3	0,29	0,27	0,19					
0,78					0,21	0,28	0,3	0,31	0,31	0,31	0,3	0,28	0,21					
0,77					0,23	0,29	0,31	0,32	0,32	0,32	0,31	0,29	0,23					
0,76					0,24	0,3	0,32	0,32	0,32	0,32	0,32	0,3	0,24					
0,75					0,26	0,31	0,32	0,33	0,33	0,33	0,32	0,31	0,26					
0,74				0,1	0,27	0,32	0,33	0,34	0,34	0,34	0,33	0,32	0,27	0,1				
0,73				0,15	0,29	0,33	0,34	0,34	0,34	0,34	0,34	0,33	0,29	0,15				
0,72				0,18	0,3	0,33	0,34	0,35	0,35	0,35	0,34	0,33	0,3	0,18				
0,71				0,21	0,31	0,34	0,35	0,35	0,35	0,35	0,35	0,34	0,31	0,21				
0,7				0,24	0,32	0,35	0,36	0,36	0,36	0,36	0,36	0,35	0,32	0,24				
0,69				0,26	0,34	0,36	0,36	0,36	0,36	0,36	0,36	0,36	0,34	0,26				
0,68				0,28	0,35	0,36	0,37	0,37	0,37	0,37	0,37	0,36	0,35	0,28				
0,67				0,3	0,36	0,37	0,37	0,37	0,37	0,37	0,37	0,37	0,36	0,3				
0,66			0,05	0,31	0,37	0,38	0,38	0,38	0,38	0,38	0,38	0,38	0,37	0,31	0,05			
0,65			0,14	0,33	0,37	0,38	0,38	0,38	0,38	0,38	0,38	0,38	0,37	0,33	0,14			
0,64			0,19	0,35	0,38	0,39	0,39	0,39	0,38	0,39	0,39	0,39	0,38	0,35	0,19			
0,63			0,23	0,36	0,39	0,4	0,39	0,39	0,39	0,39	0,39	0,4	0,39	0,36	0,23			
0,62			0,26	0,37	0,4	0,4	0,4	0,39	0,39	0,39	0,4	0,4	0,4	0,37	0,26			
0,61			0,29	0,39	0,41	0,41	0,4	0,4	0,4	0,4	0,4	0,41	0,41	0,39	0,29			
0,6			0,32	0,4	0,42	0,41	0,41	0,4	0,4	0,4	0,41	0,41	0,42	0,4	0,32			
0,59			0,34	0,41	0,42	0,42	0,41	0,41	0,4	0,41	0,41	0,42	0,42	0,41	0,34			
0,58		0,05	0,36	0,42	0,43	0,42	0,42	0,41	0,41	0,41	0,42	0,42	0,43	0,42	0,36	0,05		
0,57		0,17	0,38	0,43	0,44	0,43	0,42	0,41	0,41	0,41	0,42	0,43	0,44	0,43	0,38	0,17		
0,56		0,23	0,4	0,44	0,44	0,43	0,42	0,42	0,41	0,42	0,42	0,43	0,44	0,44	0,4	0,23		
0,55		0,28	0,42	0,45	0,45	0,44	0,43	0,42	0,42	0,42	0,43	0,44	0,45	0,45	0,42	0,28		
0,54		0,32	0,44	0,46	0,46	0,44	0,43	0,42	0,42	0,42	0,43	0,44	0,46	0,46	0,44	0,32		
0,53		0,35	0,45	0,47	0,46	0,45	0,44	0,43	0,42	0,43	0,44	0,45	0,46	0,47	0,45	0,35		
0,52	0,09	0,38	0,47	0,48	0,47	0,45	0,44	0,43	0,43	0,43	0,44	0,45	0,47	0,48	0,47	0,38	0,09	
0,51	0,21	0,41	0,48	0,49	0,48	0,46	0,44	0,43	0,43	0,43	0,44	0,46	0,48	0,49	0,48	0,41	0,21	
0,5	0,28	0,44	0,49	0,5	0,48	0,46	0,45	0,44	0,43	0,44	0,45	0,46	0,48	0,5	0,49	0,44	0,28	
0,49	0,33	0,46	0,51	0,51	0,49	0,47	0,45	0,44	0,44	0,44	0,45	0,47	0,49	0,51	0,51	0,46	0,33	0,2
0,48	0,38	0,49	0,52	0,51	0,49	0,47	0,45	0,44	0,44	0,44	0,45	0,47	0,49	0,51	0,52	0,49	0,38	0,28
0,47	0,42	0,51	0,53	0,52	0,5	0,47	0,46	0,44	0,44	0,44	0,46	0,47	0,5	0,52	0,53	0,51	0,42	0,34
0,46	0,45	0,53	0,54	0,53	0,5	0,48	0,46	0,45	0,44	0,45	0,46	0,48	0,5	0,53	0,54	0,53	0,45	0,39
0,45	0,49	0,54	0,55	0,53	0,51	0,48	0,46	0,45	0,45	0,45	0,46	0,48	0,51	0,53	0,55	0,54	0,49	0,44
0,44	0,52	0,56	0,56	0,54	0,51	0,48	0,46	0,45	0,45	0,45	0,46	0,48	0,51	0,54	0,56	0,56	0,52	0,47
0,43	0,54	0,58	0,57	0,55	0,51	0,49	0,47	0,46	0,45	0,46	0,47	0,49	0,51	0,55	0,57	0,58	0,54	0,51
0,42	0,57	0,59	0,58	0,55	0,52	0,49	0,47	0,46	0,45	0,46	0,47	0,49	0,52	0,55	0,58	0,59	0,57	0,54
0,41	0,59	0,61	0,59	0,56	0,52	0,49	0,47	0,46	0,46	0,46	0,47	0,49	0,52	0,56	0,59	0,61	0,59	0,57
0,4	0,62	0,62	0,6	0,56	0,53	0,5	0,48	0,46	0,46	0,46	0,48	0,5	0,53	0,56	0,6	0,62	0,62	0,6
0,39	0,64	0,64	0,61	0,57	0,53	0,5	0,48	0,46	0,46	0,46	0,48	0,5	0,53	0,57	0,61	0,64	0,64	0,63
0,38	0,66	0,65	0,62	0,58	0,54	0,5	0,48	0,47	0,46	0,47	0,48	0,5	0,54	0,58	0,62	0,65	0,66	0,65
0,37	0,68	0,66	0,63	0,58	0,54	0,51	0,48	0,47	0,46	0,47	0,48	0,51	0,54	0,58	0,63	0,66	0,68	0,67
0,36	0,69	0,68	0,63	0,59	0,54	0,51	0,48	0,47	0,47	0,47	0,48	0,51	0,54	0,59	0,63	0,68	0,69	0,69
0,35	0,71	0,69	0,64	0,59	0,55	0,51	0,49	0,47	0,47	0,47	0,49	0,51	0,55	0,59	0,64	0,69	0,71	0,71
0,34	0,73	0,7	0,65	0,6	0,55	0,51	0,49	0,47	0,47	0,47	0,49	0,51	0,55	0,6	0,65	0,7	0,73	0,73

0,33	0,74	0,71	0,66	0,6	0,55	0,52	0,49	0,48	0,47	0,48	0,49	0,52	0,55	0,6	0,66	0,71	0,74	0,75
0,32	0,76	0,72	0,66	0,6	0,56	0,52	0,49	0,48	0,47	0,48	0,49	0,52	0,56	0,6	0,66	0,72	0,76	0,77
0,31	0,77	0,73	0,67	0,61	0,56	0,52	0,5	0,48	0,48	0,48	0,5	0,52	0,56	0,61	0,67	0,73	0,77	0,78
0,3	0,78	0,74	0,67	0,61	0,56	0,52	0,5	0,48	0,48	0,48	0,5	0,52	0,56	0,61	0,67	0,74	0,78	0,8
0,29	0,8	0,75	0,68	0,62	0,56	0,53	0,5	0,48	0,48	0,48	0,5	0,53	0,56	0,62	0,68	0,75	0,8	0,81
0,28	0,81	0,75	0,68	0,62	0,57	0,53	0,5	0,49	0,48	0,49	0,5	0,53	0,57	0,62	0,68	0,75	0,81	0,83
0,27	0,82	0,76	0,69	0,62	0,57	0,53	0,5	0,49	0,48	0,49	0,5	0,53	0,57	0,62	0,69	0,76	0,82	0,84
0,26	0,83	0,77	0,7	0,63	0,57	0,53	0,5	0,49	0,48	0,49	0,5	0,53	0,57	0,63	0,7	0,77	0,83	0,85
0,25	0,84	0,78	0,7	0,63	0,57	0,53	0,51	0,49	0,48	0,49	0,51	0,53	0,57	0,63	0,7	0,78	0,84	0,87
0,24	0,85	0,78	0,7	0,63	0,58	0,53	0,51	0,49	0,49	0,49	0,51	0,53	0,58	0,63	0,7	0,78	0,85	0,88
0,23	0,86	0,79	0,71	0,64	0,58	0,54	0,51	0,49	0,49	0,49	0,51	0,54	0,58	0,64	0,71	0,79	0,86	0,89
0,22	0,87	0,8	0,71	0,64	0,58	0,54	0,51	0,49	0,49	0,49	0,51	0,54	0,58	0,64	0,71	0,8	0,87	0,9
0,21	0,88	0,8	0,72	0,64	0,58	0,54	0,51	0,49	0,49	0,49	0,51	0,54	0,58	0,64	0,72	0,8	0,88	0,91
0,2	0,88	0,81	0,72	0,64	0,58	0,54	0,51	0,5	0,49	0,5	0,51	0,54	0,58	0,64	0,72	0,81	0,88	0,92
0,19	0,89	0,81	0,72	0,65	0,59	0,54	0,51	0,5	0,49	0,5	0,51	0,54	0,59	0,65	0,72	0,81	0,89	0,92
0,18	0,9	0,82	0,73	0,65	0,59	0,54	0,51	0,5	0,49	0,5	0,51	0,54	0,59	0,65	0,73	0,82	0,9	0,93
0,17	0,91	0,82	0,73	0,65	0,59	0,54	0,52	0,5	0,49	0,5	0,52	0,54	0,59	0,65	0,73	0,82	0,91	0,94
0,16	0,91	0,83	0,73	0,65	0,59	0,55	0,52	0,5	0,49	0,5	0,52	0,55	0,59	0,65	0,73	0,83	0,91	0,95
0,15	0,92	0,83	0,74	0,65	0,59	0,55	0,52	0,5	0,49	0,5	0,52	0,55	0,59	0,65	0,74	0,83	0,92	0,95
0,14	0,92	0,84	0,74	0,66	0,59	0,55	0,52	0,5	0,5	0,5	0,52	0,55	0,59	0,66	0,74	0,84	0,92	0,96
0,13	0,93	0,84	0,74	0,66	0,59	0,55	0,52	0,5	0,5	0,5	0,52	0,55	0,59	0,66	0,74	0,84	0,93	0,97
0,12	0,93	0,84	0,74	0,66	0,6	0,55	0,52	0,5	0,5	0,5	0,52	0,55	0,6	0,66	0,74	0,84	0,93	0,97
0,11	0,94	0,84	0,75	0,66	0,6	0,55	0,52	0,5	0,5	0,5	0,52	0,55	0,6	0,66	0,75	0,84	0,94	0,98
0,1	0,94	0,85	0,75	0,66	0,6	0,55	0,52	0,5	0,5	0,5	0,52	0,55	0,6	0,66	0,75	0,85	0,94	0,98
0,09	0,94	0,85	0,75	0,66	0,6	0,55	0,52	0,5	0,5	0,5	0,52	0,55	0,6	0,66	0,75	0,85	0,94	0,98
0,08	0,95	0,85	0,75	0,66	0,6	0,55	0,52	0,5	0,5	0,5	0,52	0,55	0,6	0,66	0,75	0,85	0,95	0,99
0,07	0,95	0,85	0,75	0,67	0,6	0,55	0,52	0,5	0,5	0,5	0,52	0,55	0,6	0,67	0,75	0,85	0,95	0,99
0,06	0,95	0,86	0,75	0,67	0,6	0,55	0,52	0,5	0,5	0,5	0,52	0,55	0,6	0,67	0,75	0,86	0,95	0,99
0,05	0,95	0,86	0,75	0,67	0,6	0,55	0,52	0,51	0,5	0,51	0,52	0,55	0,6	0,67	0,75	0,86	0,95	0,99
0,04	0,95	0,86	0,75	0,67	0,6	0,55	0,52	0,51	0,5	0,51	0,52	0,55	0,6	0,67	0,75	0,86	0,95	1
0,03	0,96	0,86	0,76	0,67	0,6	0,55	0,52	0,51	0,5	0,51	0,52	0,55	0,6	0,67	0,76	0,86	0,96	1
0,02	0,96	0,86	0,76	0,67	0,6	0,55	0,52	0,51	0,5	0,51	0,52	0,55	0,6	0,67	0,76	0,86	0,96	1
0,01	0,96	0,86	0,76	0,67	0,6	0,55	0,52	0,51	0,5	0,51	0,52	0,55	0,6	0,67	0,76	0,86	0,96	1
-0	0,96	0,86	0,76	0,67	0,6	0,55	0,52	0,51	0,5	0,51	0,52	0,55	0,6	0,67	0,76	0,86	0,96	1
-0,01	0,96	0,86	0,76	0,67	0,6	0,55	0,52	0,51	0,5	0,51	0,52	0,55	0,6	0,67	0,76	0,86	0,96	1
-0,02	0,96	0,86	0,76	0,67	0,6	0,55	0,52	0,51	0,5	0,51	0,52	0,55	0,6	0,67	0,76	0,86	0,96	1
-0,03	0,96	0,86	0,76	0,67	0,6	0,55	0,52	0,51	0,5	0,51	0,52	0,55	0,6	0,67	0,76	0,86	0,96	1
-0,04	0,95	0,86	0,75	0,67	0,6	0,55	0,52	0,51	0,5	0,51	0,52	0,55	0,6	0,67	0,75	0,86	0,95	1
-0,05	0,95	0,86	0,75	0,67	0,6	0,55	0,52	0,51	0,5	0,51	0,52	0,55	0,6	0,67	0,75	0,86	0,95	0,99
-0,06	0,95	0,86	0,75	0,67	0,6	0,55	0,52	0,5	0,5	0,5	0,52	0,55	0,6	0,67	0,75	0,86	0,95	0,99
-0,07	0,95	0,85	0,75	0,67	0,6	0,55	0,52	0,5	0,5	0,5	0,52	0,55	0,6	0,67	0,75	0,85	0,95	0,99
-0,08	0,95	0,85	0,75	0,66	0,6	0,55	0,52	0,5	0,5	0,5	0,52	0,55	0,6	0,66	0,75	0,85	0,95	0,99
-0,09	0,94	0,85	0,75	0,66	0,6	0,55	0,52	0,5	0,5	0,5	0,52	0,55	0,6	0,66	0,75	0,85	0,94	0,98
-0,1	0,94	0,85	0,75	0,66	0,6	0,55	0,52	0,5	0,5	0,5	0,52	0,55	0,6	0,66	0,75	0,85	0,94	0,98
-0,11	0,94	0,84	0,75	0,66	0,6	0,55	0,52	0,5	0,5	0,5	0,52	0,55	0,6	0,66	0,75	0,84	0,94	0,98
-0,12	0,93	0,84	0,74	0,66	0,6	0,55	0,52	0,5	0,5	0,5	0,52	0,55	0,6	0,66	0,74	0,84	0,93	0,97
-0,13	0,93	0,84	0,74	0,66	0,59	0,55	0,52	0,5	0,5	0,5	0,52	0,55	0,59	0,66	0,74	0,84	0,93	0,97
-0,14	0,92	0,84	0,74	0,66	0,59	0,55	0,52	0,5	0,5	0,5	0,52	0,55	0,59	0,66	0,74	0,84	0,92	0,96
-0,15	0,92	0,83	0,74	0,65	0,59	0,55	0,52	0,5	0,49	0,5	0,52	0,55	0,59	0,65	0,74	0,83	0,92	0,95
-0,16	0,91	0,83	0,73	0,65	0,59	0,55	0,52	0,5	0,49	0,5	0,52	0,55	0,59	0,65	0,73	0,83	0,91	0,95
-0,17	0,91	0,82	0,73	0,65	0,59	0,54	0,52	0,5	0,49	0,5	0,52	0,54	0,59	0,65	0,73	0,82	0,91	0,94
-0,18	0,9	0,82	0,73	0,65	0,59	0,54	0,51	0,5	0,49	0,5	0,51	0,54	0,59	0,65	0,73	0,82	0,9	0,93
-0,19	0,89	0,81	0,72	0,65	0,59	0,54	0,51	0,5	0,49	0,5	0,51	0,54	0,59	0,65	0,72	0,81	0,89	0,92
-0,2	0,88	0,81	0,72	0,64	0,58	0,54	0,51	0,5	0,49	0,5	0,51	0,54	0,58	0,64	0,72	0,81	0,88	0,92
-0,21	0,88	0,8	0,72	0,64	0,58	0,54	0,51	0,49	0,49	0,49	0,51	0,54	0,58	0,64	0,72	0,8	0,88	0,91
-0,22	0,87	0,8	0,71	0,64	0,58	0,54	0,51	0,49	0,49	0,49	0,51	0,54	0,58	0,64	0,71	0,8	0,87	0,9
-0,23	0,86	0,79	0,71	0,64	0,58	0,54	0,51	0,49	0,49	0,49	0,51	0,54	0,58	0,64	0,71	0,79	0,86	0,89
-0,24	0,85	0,78	0,7	0,63	0,58	0,53	0,51	0,49	0,49	0,49	0,51	0,53	0,58	0,63	0,7	0,78	0,85	0,88
-0,25	0,84	0,78	0,7	0,63	0,57	0,53	0,51	0,49	0,48	0,49	0,51	0,53	0,57	0,63	0,7	0,78	0,84	0,87
-0,26	0,83	0,77	0,7	0,63	0,57	0,53	0,5	0,49	0,48	0,49	0,5	0,53	0,57	0,63	0,7	0,77	0,83	0,85
-0,27	0,82	0,76	0,69	0,62	0,57	0,53	0,5	0,49	0,48	0,49	0,5	0,53	0,57	0,62	0,69	0,76	0,82	0,84
-0,28	0,81	0,75	0,68	0,62	0,57	0,53	0,5	0,49	0,48	0,49	0,5	0,53	0,57	0,62	0,68	0,75	0,81	0,83
-0,29	0,8	0,75	0,68	0,62	0,56	0,53	0,5	0,48	0,48	0,48	0,5	0,53	0,56	0,62	0,68	0,75	0,8	0,81
-0,3	0,78	0,74	0,67	0,61	0,56	0,52	0,5	0,48	0,48	0,48	0,5	0,52	0,56	0,61	0,67	0,74	0,78	0,8
-0,31	0,77	0,73	0,67	0,61	0,56	0,52	0,5	0,48	0,48	0,48	0,5	0,52	0,56	0,61	0,67	0,73	0,77	0,78
-0,32	0,76	0,72	0,66	0,6	0,56	0,52	0,49	0,48	0,47	0,48	0,49	0,52	0,56	0,6	0,66	0,72	0,76	0,77
-0,33	0,74	0,71	0,66	0,6	0,55	0,52	0,49	0,48	0,47	0,48	0,49	0,52	0,56	0,6	0,66	0,71	0,74	0,75
-0,34	0,73	0,7	0,65	0,6	0,55	0,51	0,49	0,47	0,47	0,47	0,49	0,51	0,55	0,6	0,65	0,7	0,73	0,73
-0,35	0,71	0,69	0,64	0,59	0,55	0,51	0,49	0,47	0,47	0,47	0,49	0,51	0,55	0,59	0,64	0,69	0,71	0,71
-0,36	0,69	0,68	0,63	0,59	0,54	0,51	0,48	0,47	0,47	0,47	0,48	0,51	0,54	0,59	0,63	0,68	0,69	0,69
-0,37	0,68	0,66	0,63	0,58	0,54	0,51	0,48	0,47	0,46	0,47	0,48	0,51	0,54	0,58	0,63	0,66	0,68	0,67
-0,38	0,66	0,65	0,62	0,58	0,54	0,5	0,48	0,47	0,46	0,47	0,48	0,5	0,54	0,58	0,62	0,65	0,66	0,65
-0,39	0,64	0,64	0,61	0,57	0,53	0,5	0,48	0,46	0,46	0,46	0,48	0,5	0,53	0,57	0,61	0,64	0,64	0,63
-0,4	0,62	0,62	0,6	0,56	0,53	0,5	0,48	0,46	0,46	0,46	0,48	0,5	0,53	0,56	0,6	0,62	0,62	0,6
-0,41	0,59	0,61	0,59	0,56	0,52	0,49	0,47	0,46	0,46	0,46	0,47	0,49	0,52	0,56	0,59	0,61	0,59	0,57
-0,42	0,57	0,59	0,58	0,55	0,52	0,49	0,47	0,46	0,45	0,46	0,47	0,49	0,52	0,55	0,58	0,59	0,57	0,54

$\bar h_{\beta/x}\rightarrow$	0,75	0,67	0,59	0,53	0,47	0,43	0,41	0,4	0,39	0,4	0,41	0,43	0,47	0,53	0,59	0,67	0,75	0,79
-0,43	0,54	0,58	0,57	0,55	0,51	0,49	0,47	0,46	0,45	0,46	0,47	0,49	0,51	0,55	0,57	0,58	0,54	0,51
-0,44	0,52	0,56	0,56	0,54	0,51	0,48	0,46	0,45	0,45	0,45	0,46	0,48	0,51	0,54	0,56	0,56	0,52	0,47
-0,45	0,49	0,54	0,55	0,53	0,51	0,48	0,46	0,45	0,45	0,45	0,46	0,48	0,51	0,53	0,55	0,54	0,49	0,44
-0,46	0,45	0,53	0,54	0,53	0,5	0,48	0,46	0,45	0,44	0,45	0,46	0,48	0,5	0,53	0,54	0,53	0,45	0,39
-0,47	0,42	0,51	0,53	0,52	0,5	0,47	0,46	0,44	0,44	0,44	0,46	0,47	0,5	0,52	0,53	0,51	0,42	0,34
-0,48	0,38	0,49	0,52	0,51	0,49	0,47	0,45	0,44	0,44	0,44	0,45	0,47	0,49	0,51	0,52	0,49	0,38	0,28
-0,49	0,33	0,46	0,51	0,51	0,49	0,47	0,45	0,44	0,44	0,44	0,45	0,47	0,49	0,51	0,51	0,46	0,33	0,2
-0,5	0,28	0,44	0,49	0,5	0,48	0,46	0,45	0,44	0,43	0,44	0,45	0,46	0,48	0,5	0,49	0,44	0,28	
-0,51	0,21	0,41	0,48	0,49	0,48	0,46	0,44	0,43	0,43	0,43	0,44	0,46	0,48	0,49	0,48	0,41	0,21	
-0,52	0,09	0,38	0,47	0,48	0,47	0,45	0,44	0,43	0,43	0,43	0,44	0,45	0,47	0,48	0,47	0,38	0,09	
-0,53		0,35	0,45	0,47	0,46	0,45	0,44	0,43	0,42	0,43	0,44	0,45	0,46	0,47	0,45	0,35		
-0,54		0,32	0,44	0,46	0,46	0,44	0,43	0,42	0,42	0,42	0,43	0,44	0,46	0,46	0,44	0,32		
-0,55		0,28	0,42	0,45	0,45	0,44	0,43	0,42	0,42	0,42	0,43	0,44	0,45	0,45	0,42	0,28		
-0,56		0,23	0,4	0,44	0,44	0,43	0,42	0,42	0,41	0,42	0,42	0,43	0,44	0,44	0,4	0,23		
-0,57		0,17	0,38	0,43	0,44	0,43	0,42	0,41	0,41	0,41	0,42	0,43	0,44	0,43	0,38	0,17		
-0,58		0,05	0,36	0,42	0,43	0,42	0,42	0,41	0,41	0,41	0,42	0,42	0,43	0,42	0,36	0,05		
-0,59			0,34	0,41	0,42	0,42	0,41	0,41	0,4	0,41	0,41	0,42	0,42	0,41	0,34			
-0,6			0,32	0,4	0,42	0,41	0,41	0,4	0,4	0,4	0,41	0,41	0,42	0,4	0,32			
-0,61			0,29	0,39	0,41	0,41	0,4	0,4	0,4	0,4	0,4	0,41	0,41	0,39	0,29			
-0,62			0,26	0,37	0,4	0,4	0,4	0,39	0,39	0,39	0,4	0,4	0,4	0,37	0,26			
-0,63			0,23	0,36	0,39	0,4	0,39	0,39	0,39	0,39	0,39	0,4	0,39	0,36	0,23			
-0,64			0,19	0,35	0,38	0,39	0,39	0,39	0,38	0,39	0,39	0,39	0,38	0,35	0,19			
-0,65			0,14	0,33	0,37	0,38	0,38	0,38	0,38	0,38	0,38	0,38	0,37	0,33	0,14			
-0,66			0,05	0,31	0,37	0,38	0,38	0,38	0,38	0,38	0,38	0,38	0,37	0,31	0,05			
-0,67				0,3	0,36	0,37	0,37	0,37	0,37	0,37	0,37	0,37	0,36	0,3				
-0,68				0,28	0,35	0,36	0,37	0,37	0,37	0,37	0,37	0,36	0,35	0,28				
-0,69				0,26	0,34	0,36	0,36	0,36	0,36	0,36	0,36	0,36	0,34	0,26				
-0,7				0,24	0,32	0,35	0,36	0,36	0,36	0,36	0,36	0,35	0,32	0,24				
-0,71				0,21	0,31	0,34	0,35	0,35	0,35	0,35	0,35	0,34	0,31	0,21				
-0,72				0,18	0,3	0,33	0,34	0,35	0,35	0,35	0,34	0,33	0,3	0,18				
-0,73				0,15	0,29	0,33	0,34	0,34	0,34	0,34	0,34	0,33	0,29	0,15				
-0,74				0,1	0,27	0,32	0,33	0,34	0,34	0,34	0,33	0,32	0,27	0,1				
-0,75					0,26	0,31	0,32	0,33	0,33	0,33	0,32	0,31	0,26					
-0,76					0,24	0,3	0,32	0,32	0,32	0,32	0,32	0,3	0,24					
-0,77					0,23	0,29	0,31	0,32	0,32	0,32	0,31	0,29	0,23					
-0,78					0,21	0,28	0,3	0,31	0,31	0,31	0,3	0,28	0,21					
-0,79					0,19	0,27	0,29	0,3	0,31	0,3	0,29	0,27	0,19					
-0,8					0,16	0,26	0,29	0,3	0,3	0,3	0,29	0,26	0,16					
-0,81					0,13	0,24	0,28	0,29	0,29	0,29	0,28	0,24	0,13					
-0,82					0,1	0,23	0,27	0,28	0,29	0,28	0,27	0,23	0,1					
-0,83					0,03	0,22	0,26	0,27	0,28	0,27	0,26	0,22	0,03					
-0,84						0,2	0,25	0,27	0,27	0,27	0,25	0,2						
-0,85						0,18	0,24	0,26	0,26	0,26	0,24	0,18						
-0,86						0,17	0,23	0,25	0,26	0,25	0,23	0,17						
-0,87						0,15	0,22	0,24	0,25	0,24	0,22	0,15						
-0,88						0,12	0,2	0,23	0,24	0,23	0,2	0,12						
-0,89						0,09	0,19	0,22	0,23	0,22	0,19	0,09						
-0,9						0,03	0,18	0,21	0,22	0,21	0,18	0,03						
-0,91							0,16	0,2	0,21	0,2	0,16							
-0,92							0,14	0,19	0,2	0,19	0,14							
-0,93							0,12	0,17	0,18	0,17	0,12							
-0,94							0,09	0,16	0,17	0,16	0,09							
-0,95							0,05	0,14	0,16	0,14	0,05							
-0,96								0,12	0,14	0,12								
-0,97								0,1	0,12	0,1								
-0,98								0,07	0,1	0,07								
-0,99									0,07									
-1																		

$$\sum_{\Delta\vec h_x} H_{\beta i/x} = \text{78,6} \quad \text{78,6} \quad \text{78,6} \quad \text{78,5} \quad \text{78,5} \quad \text{78,5} \quad \text{78,5} \quad \text{78,5} \quad \text{78,5} \quad \text{78,5} \quad \text{78,5} \quad \text{78,5} \quad \text{78,5} \quad \text{78,5} \quad \text{78,6} \quad \text{78,6} \quad \text{78,6} \quad \text{78,5}$$

$$n_{\beta,\Delta\vec h_x} = \text{105} \quad \text{117} \quad \text{133} \quad \text{149} \quad \text{167} \quad \text{181} \quad \text{191} \quad \text{197} \quad \text{199} \quad \text{197} \quad \text{191} \quad \text{181} \quad \text{167} \quad \text{149} \quad \text{133} \quad \text{117} \quad \text{105} \quad \text{99}$$

$$\bar h_{\beta/x} = \text{0,75} \quad \text{0,67} \quad \text{0,59} \quad \text{0,53} \quad \text{0,47} \quad \text{0,43} \quad \text{0,41} \quad \text{0,4} \quad \text{0,39} \quad \text{0,4} \quad \text{0,41} \quad \text{0,43} \quad \text{0,47} \quad \text{0,53} \quad \text{0,59} \quad \text{0,67} \quad \text{0,75} \quad \text{0,79}$$

$$d_{\beta/x} = \text{0,96} \quad \text{0,86} \quad \text{0,76} \quad \text{0,67} \quad \text{0,6} \quad \text{0,55} \quad \text{0,52} \quad \text{0,51} \quad \text{0,5} \quad \text{0,51} \quad \text{0,52} \quad \text{0,55} \quad \text{0,6} \quad \text{0,67} \quad \text{0,76} \quad \text{0,86} \quad \text{0,96} \quad \text{1}$$

$$\bar h_\beta / d_\beta = \text{0,78} \quad \text{0,78} \quad \text{0,78} \quad \text{0,79} \quad \text{0,78} \quad \text{0,78} \quad \text{0,79} \quad \text{0,79} \quad \text{0,79} \quad \text{0,79} \quad \text{0,79} \quad \text{0,78} \quad \text{0,78} \quad \text{0,79} \quad \text{0,78} \quad \text{0,78} \quad \text{0,78} \quad \text{0,79}$$

$$\bar{\bar h}_{k\beta/x(\Delta\vec h_x)} = 0,54; \quad \bar{\bar h}_{k/x,\Delta\vec h_x} = 0,51; \quad \bar{\bar h}_{k\beta/x(\Delta\vec h_x)}/\bar d_{k/x} = 0,784 \rightarrow \pi/4 = 0,785398$$

$$k = 0,3$$

k \ β	0,17	0,35	0,52	0,7	0,87	1,05	1,22	1,4	1,57	1,75	1,92	2,09	2,27	2,44	2,62	2,79	2,97	3,14
0,3 Δh/a																		
1																		
0,99										0,04								
0,98									0,03	0,06	0,03							
0,97									0,05	0,07	0,05							
0,96									0,07	0,08	0,07							
0,95									0,08	0,09	0,08							
0,94								0,03	0,09	0,1	0,09	0,03						
0,93								0,06	0,1	0,11	0,1	0,06						
0,92								0,07	0,11	0,12	0,11	0,07						
0,91								0,09	0,12	0,12	0,12	0,09						
0,9								0,1	0,12	0,13	0,12	0,1						
0,89								0,11	0,13	0,14	0,13	0,11						
0,88								0,12	0,14	0,14	0,14	0,12						
0,87							0,05	0,12	0,14	0,15	0,14	0,12	0,05					
0,86							0,07	0,13	0,15	0,15	0,15	0,13	0,07					
0,85							0,09	0,14	0,15	0,16	0,15	0,14	0,09					
0,84							0,1	0,15	0,16	0,16	0,16	0,15	0,1					
0,83							0,11	0,15	0,16	0,17	0,16	0,15	0,11					
0,82							0,12	0,16	0,17	0,17	0,17	0,16	0,12					
0,81							0,13	0,16	0,17	0,18	0,17	0,16	0,13					
0,8							0,14	0,17	0,18	0,18	0,18	0,17	0,14					
0,79							0,15	0,17	0,18	0,18	0,18	0,17	0,15					
0,78						0,06	0,16	0,18	0,19	0,19	0,19	0,18	0,16	0,06				
0,77						0,08	0,16	0,18	0,19	0,19	0,19	0,18	0,16	0,08				
0,76						0,1	0,17	0,19	0,19	0,19	0,19	0,19	0,17	0,1				
0,75						0,12	0,18	0,19	0,2	0,2	0,2	0,19	0,18	0,12				
0,74						0,13	0,18	0,2	0,2	0,2	0,2	0,2	0,18	0,13				
0,73						0,15	0,19	0,2	0,2	0,21	0,2	0,2	0,19	0,15				
0,72						0,16	0,2	0,21	0,21	0,21	0,21	0,21	0,2	0,16				
0,71						0,17	0,2	0,21	0,21	0,21	0,21	0,21	0,2	0,17				
0,7						0,18	0,21	0,21	0,21	0,21	0,21	0,21	0,21	0,18				
0,69						0,18	0,21	0,22	0,22	0,22	0,22	0,22	0,21	0,18				
0,68					0,04	0,19	0,22	0,22	0,22	0,22	0,22	0,22	0,22	0,19	0,04			
0,67					0,08	0,2	0,22	0,22	0,22	0,22	0,22	0,22	0,22	0,2	0,08			
0,66					0,11	0,21	0,23	0,23	0,23	0,23	0,23	0,23	0,23	0,21	0,11			
0,65					0,13	0,22	0,23	0,23	0,23	0,23	0,23	0,23	0,23	0,22	0,13			
0,64					0,15	0,22	0,23	0,23	0,23	0,23	0,23	0,23	0,23	0,22	0,15			
0,63					0,17	0,23	0,24	0,24	0,23	0,23	0,23	0,24	0,24	0,23	0,17			
0,62					0,18	0,24	0,24	0,24	0,24	0,24	0,24	0,24	0,24	0,24	0,18			
0,61					0,2	0,24	0,25	0,24	0,24	0,24	0,24	0,24	0,25	0,24	0,2			
0,6					0,21	0,25	0,25	0,25	0,24	0,24	0,24	0,25	0,25	0,25	0,21			
0,59					0,22	0,25	0,25	0,25	0,24	0,24	0,24	0,25	0,25	0,25	0,22			
0,58					0,23	0,26	0,26	0,25	0,25	0,24	0,25	0,25	0,26	0,26	0,23			
0,57					0,24	0,26	0,26	0,25	0,25	0,25	0,25	0,25	0,26	0,26	0,24			
0,56				0,06	0,25	0,27	0,26	0,25	0,25	0,25	0,25	0,26	0,26	0,27	0,25	0,06		
0,55				0,12	0,26	0,27	0,27	0,26	0,25	0,25	0,25	0,26	0,27	0,27	0,26	0,12		
0,54				0,15	0,27	0,28	0,27	0,26	0,25	0,25	0,25	0,26	0,27	0,28	0,27	0,15		
0,53				0,18	0,28	0,28	0,27	0,26	0,26	0,25	0,26	0,26	0,27	0,28	0,28	0,18		
0,52				0,21	0,28	0,29	0,28	0,27	0,26	0,26	0,26	0,27	0,28	0,29	0,28	0,21		
0,51				0,23	0,29	0,29	0,28	0,27	0,26	0,26	0,26	0,27	0,28	0,29	0,29	0,23		
0,5				0,25	0,3	0,29	0,28	0,27	0,26	0,26	0,26	0,27	0,28	0,29	0,3	0,25		
0,49				0,26	0,31	0,3	0,28	0,27	0,26	0,26	0,26	0,27	0,28	0,3	0,31	0,26		
0,48				0,28	0,31	0,3	0,29	0,27	0,27	0,26	0,27	0,27	0,29	0,3	0,31	0,28		
0,47				0,29	0,32	0,31	0,29	0,28	0,27	0,26	0,27	0,28	0,29	0,31	0,32	0,29		
0,46				0,31	0,32	0,31	0,29	0,28	0,27	0,27	0,27	0,28	0,29	0,31	0,32	0,31		
0,45				0,32	0,33	0,31	0,29	0,28	0,27	0,27	0,27	0,28	0,29	0,31	0,33	0,32		
0,44			0,08	0,33	0,34	0,32	0,3	0,28	0,27	0,27	0,27	0,28	0,3	0,32	0,34	0,33	0,08	
0,43			0,16	0,34	0,34	0,32	0,3	0,28	0,27	0,27	0,27	0,28	0,3	0,32	0,34	0,34	0,16	
0,42			0,22	0,35	0,35	0,32	0,3	0,28	0,28	0,28	0,28	0,28	0,3	0,32	0,35	0,35	0,22	
0,41			0,26	0,37	0,35	0,32	0,3	0,29	0,28	0,27	0,28	0,29	0,3	0,32	0,35	0,37	0,26	
0,4			0,29	0,37	0,36	0,33	0,3	0,29	0,28	0,27	0,28	0,29	0,3	0,33	0,36	0,37	0,29	
0,39			0,32	0,38	0,36	0,33	0,31	0,29	0,28	0,28	0,28	0,29	0,31	0,33	0,36	0,38	0,32	
0,38			0,35	0,39	0,37	0,33	0,31	0,29	0,28	0,28	0,28	0,29	0,31	0,33	0,37	0,39	0,35	
0,37			0,37	0,4	0,37	0,34	0,31	0,29	0,28	0,28	0,28	0,29	0,31	0,34	0,37	0,4	0,37	
0,36			0,39	0,41	0,37	0,34	0,31	0,29	0,28	0,28	0,28	0,29	0,31	0,34	0,37	0,41	0,39	
0,35			0,42	0,42	0,38	0,34	0,31	0,29	0,28	0,28	0,28	0,29	0,31	0,34	0,38	0,42	0,42	
0,34	0,11	0,43	0,42	0,38	0,34	0,31	0,3	0,29	0,28	0,29	0,3	0,31	0,34	0,38	0,42	0,43	0,11	

42

0,33	0,24	0,45	0,43	0,38	0,35	0,32	0,3	0,29	0,28	0,29	0,3	0,32	0,35	0,38	0,43	0,45	0,24	
0,32	0,31	0,47	0,44	0,39	0,35	0,32	0,3	0,29	0,28	0,29	0,3	0,32	0,35	0,39	0,44	0,47	0,31	
0,31	0,37	0,48	0,44	0,39	0,35	0,32	0,3	0,29	0,29	0,29	0,3	0,32	0,35	0,39	0,44	0,48	0,37	
0,3	0,42	0,5	0,45	0,39	0,35	0,32	0,3	0,29	0,29	0,29	0,3	0,32	0,35	0,39	0,45	0,5	0,42	
0,29	0,47	0,51	0,46	0,4	0,35	0,32	0,3	0,29	0,29	0,29	0,3	0,32	0,35	0,4	0,46	0,51	0,47	0,26
0,28	0,5	0,52	0,46	0,4	0,36	0,32	0,3	0,29	0,29	0,29	0,3	0,32	0,36	0,4	0,46	0,52	0,5	0,36
0,27	0,54	0,54	0,47	0,4	0,36	0,32	0,3	0,29	0,29	0,29	0,3	0,32	0,36	0,4	0,47	0,54	0,54	0,44
0,26	0,57	0,55	0,47	0,41	0,36	0,33	0,31	0,29	0,29	0,29	0,31	0,33	0,36	0,41	0,47	0,55	0,57	0,5
0,25	0,6	0,56	0,48	0,41	0,36	0,33	0,31	0,29	0,29	0,29	0,31	0,33	0,36	0,41	0,48	0,56	0,6	0,55
0,24	0,62	0,57	0,48	0,41	0,36	0,33	0,31	0,3	0,29	0,3	0,31	0,33	0,36	0,41	0,48	0,57	0,62	0,6
0,23	0,65	0,58	0,49	0,41	0,36	0,33	0,31	0,3	0,29	0,3	0,31	0,33	0,36	0,41	0,49	0,58	0,65	0,64
0,22	0,67	0,59	0,49	0,42	0,36	0,33	0,31	0,3	0,29	0,3	0,31	0,33	0,36	0,42	0,49	0,59	0,67	0,68
0,21	0,69	0,6	0,49	0,42	0,37	0,33	0,31	0,3	0,29	0,3	0,31	0,33	0,37	0,42	0,49	0,6	0,69	0,71
0,2	0,71	0,6	0,5	0,42	0,37	0,33	0,31	0,3	0,29	0,3	0,31	0,33	0,37	0,42	0,5	0,6	0,71	0,75
0,19	0,73	0,61	0,5	0,42	0,37	0,33	0,31	0,3	0,29	0,3	0,31	0,33	0,37	0,42	0,5	0,61	0,73	0,77
0,18	0,74	0,62	0,5	0,42	0,37	0,33	0,31	0,3	0,3	0,3	0,31	0,33	0,37	0,42	0,5	0,62	0,74	0,8
0,17	0,76	0,63	0,51	0,43	0,37	0,33	0,31	0,3	0,3	0,3	0,31	0,33	0,37	0,43	0,51	0,63	0,76	0,82
0,16	0,77	0,63	0,51	0,43	0,37	0,34	0,31	0,3	0,3	0,3	0,31	0,34	0,37	0,43	0,51	0,63	0,77	0,85
0,15	0,79	0,64	0,51	0,43	0,37	0,34	0,31	0,3	0,3	0,3	0,31	0,34	0,37	0,43	0,51	0,64	0,79	0,87
0,14	0,8	0,64	0,52	0,43	0,37	0,34	0,31	0,3	0,3	0,3	0,31	0,34	0,37	0,43	0,52	0,64	0,8	0,88
0,13	0,81	0,65	0,52	0,43	0,37	0,34	0,31	0,3	0,3	0,3	0,31	0,34	0,37	0,43	0,52	0,65	0,81	0,9
0,12	0,82	0,65	0,52	0,43	0,38	0,34	0,31	0,3	0,3	0,3	0,31	0,34	0,38	0,43	0,52	0,65	0,82	0,92
0,11	0,83	0,66	0,52	0,43	0,38	0,34	0,32	0,3	0,3	0,3	0,32	0,34	0,38	0,43	0,52	0,66	0,83	0,93
0,1	0,84	0,66	0,52	0,43	0,38	0,34	0,32	0,3	0,3	0,3	0,32	0,34	0,38	0,43	0,52	0,66	0,84	0,94
0,09	0,84	0,66	0,53	0,44	0,38	0,34	0,32	0,3	0,3	0,3	0,32	0,34	0,38	0,44	0,53	0,66	0,84	0,95
0,08	0,85	0,67	0,53	0,44	0,38	0,34	0,32	0,3	0,3	0,3	0,32	0,34	0,38	0,44	0,53	0,67	0,85	0,96
0,07	0,86	0,67	0,53	0,44	0,38	0,34	0,32	0,3	0,3	0,3	0,32	0,34	0,38	0,44	0,53	0,67	0,86	0,97
0,06	0,86	0,67	0,53	0,44	0,38	0,34	0,32	0,3	0,3	0,3	0,32	0,34	0,38	0,44	0,53	0,67	0,86	0,98
0,05	0,87	0,67	0,53	0,44	0,38	0,34	0,32	0,3	0,3	0,3	0,32	0,34	0,38	0,44	0,53	0,67	0,87	0,99
0,04	0,87	0,67	0,53	0,44	0,38	0,34	0,32	0,3	0,3	0,3	0,32	0,34	0,38	0,44	0,53	0,67	0,87	0,99
0,03	0,87	0,68	0,53	0,44	0,38	0,34	0,32	0,3	0,3	0,3	0,32	0,34	0,38	0,44	0,53	0,68	0,87	0,99
0,02	0,87	0,68	0,53	0,44	0,38	0,34	0,32	0,3	0,3	0,3	0,32	0,34	0,38	0,44	0,53	0,68	0,87	1
0,01	0,88	0,68	0,53	0,44	0,38	0,34	0,32	0,3	0,3	0,3	0,32	0,34	0,38	0,44	0,53	0,68	0,88	1
-0	0,88	0,68	0,53	0,44	0,38	0,34	0,32	0,3	0,3	0,3	0,32	0,34	0,38	0,44	0,53	0,68	0,88	1
-0,01	0,88	0,68	0,53	0,44	0,38	0,34	0,32	0,3	0,3	0,3	0,32	0,34	0,38	0,44	0,53	0,68	0,88	1
-0,02	0,87	0,68	0,53	0,44	0,38	0,34	0,32	0,3	0,3	0,3	0,32	0,34	0,38	0,44	0,53	0,68	0,87	1
-0,03	0,87	0,68	0,53	0,44	0,38	0,34	0,32	0,3	0,3	0,3	0,32	0,34	0,38	0,44	0,53	0,68	0,87	0,99
-0,04	0,87	0,67	0,53	0,44	0,38	0,34	0,32	0,3	0,3	0,3	0,32	0,34	0,38	0,44	0,53	0,67	0,87	0,99
-0,05	0,87	0,67	0,53	0,44	0,38	0,34	0,32	0,3	0,3	0,3	0,32	0,34	0,38	0,44	0,53	0,67	0,87	0,99
-0,06	0,86	0,67	0,53	0,44	0,38	0,34	0,32	0,3	0,3	0,3	0,32	0,34	0,38	0,44	0,53	0,67	0,86	0,98
-0,07	0,86	0,67	0,53	0,44	0,38	0,34	0,32	0,3	0,3	0,3	0,32	0,34	0,38	0,44	0,53	0,67	0,86	0,97
-0,08	0,85	0,67	0,53	0,44	0,38	0,34	0,32	0,3	0,3	0,3	0,32	0,34	0,38	0,44	0,53	0,67	0,85	0,96
-0,09	0,84	0,66	0,53	0,44	0,38	0,34	0,32	0,3	0,3	0,3	0,32	0,34	0,38	0,44	0,53	0,66	0,84	0,95
-0,1	0,84	0,66	0,52	0,43	0,38	0,34	0,32	0,3	0,3	0,3	0,32	0,34	0,38	0,43	0,52	0,66	0,84	0,94
-0,11	0,83	0,66	0,52	0,43	0,38	0,34	0,32	0,3	0,3	0,3	0,32	0,34	0,38	0,43	0,52	0,66	0,83	0,93
-0,12	0,82	0,65	0,52	0,43	0,38	0,34	0,31	0,3	0,3	0,3	0,31	0,34	0,38	0,43	0,52	0,65	0,82	0,92
-0,13	0,81	0,65	0,52	0,43	0,37	0,34	0,31	0,3	0,3	0,3	0,31	0,34	0,37	0,43	0,52	0,65	0,81	0,9
-0,14	0,8	0,64	0,52	0,43	0,37	0,34	0,31	0,3	0,3	0,3	0,31	0,34	0,37	0,43	0,52	0,64	0,8	0,88
-0,15	0,79	0,64	0,51	0,43	0,37	0,34	0,31	0,3	0,3	0,3	0,31	0,34	0,37	0,43	0,51	0,64	0,79	0,87
-0,16	0,77	0,63	0,51	0,43	0,37	0,34	0,31	0,3	0,3	0,3	0,31	0,34	0,37	0,43	0,51	0,63	0,77	0,85
-0,17	0,76	0,63	0,51	0,43	0,37	0,33	0,31	0,3	0,3	0,3	0,31	0,33	0,37	0,43	0,51	0,63	0,76	0,82
-0,18	0,74	0,62	0,5	0,42	0,37	0,33	0,31	0,3	0,3	0,3	0,31	0,33	0,37	0,42	0,5	0,62	0,74	0,8
-0,19	0,73	0,61	0,5	0,42	0,37	0,33	0,31	0,3	0,29	0,3	0,31	0,33	0,37	0,42	0,5	0,61	0,73	0,77
-0,2	0,71	0,6	0,5	0,42	0,37	0,33	0,31	0,3	0,29	0,3	0,31	0,33	0,37	0,42	0,5	0,6	0,71	0,75
-0,21	0,69	0,6	0,49	0,42	0,37	0,33	0,31	0,3	0,29	0,3	0,31	0,33	0,37	0,42	0,49	0,6	0,69	0,71
-0,22	0,67	0,59	0,49	0,42	0,36	0,33	0,31	0,3	0,29	0,3	0,31	0,33	0,36	0,42	0,49	0,59	0,67	0,68
-0,23	0,65	0,58	0,49	0,41	0,36	0,33	0,31	0,3	0,29	0,3	0,31	0,33	0,36	0,41	0,49	0,58	0,65	0,64
-0,24	0,62	0,57	0,48	0,41	0,36	0,33	0,31	0,3	0,29	0,3	0,31	0,33	0,36	0,41	0,48	0,57	0,62	0,6
-0,25	0,6	0,56	0,48	0,41	0,36	0,33	0,31	0,29	0,29	0,29	0,31	0,33	0,36	0,41	0,48	0,56	0,6	0,55
-0,26	0,57	0,55	0,47	0,41	0,36	0,33	0,31	0,29	0,29	0,29	0,31	0,33	0,36	0,41	0,47	0,55	0,57	0,5
-0,27	0,54	0,54	0,47	0,4	0,36	0,32	0,3	0,29	0,29	0,29	0,3	0,32	0,36	0,4	0,47	0,54	0,54	0,44
-0,28	0,5	0,52	0,46	0,4	0,36	0,32	0,3	0,29	0,29	0,29	0,3	0,32	0,36	0,4	0,46	0,52	0,5	0,36
-0,29	0,47	0,51	0,46	0,4	0,35	0,32	0,3	0,29	0,29	0,29	0,3	0,32	0,35	0,4	0,46	0,51	0,47	0,26
-0,3	0,42	0,5	0,45	0,39	0,35	0,32	0,3	0,29	0,29	0,29	0,3	0,32	0,35	0,39	0,45	0,5	0,42	
-0,31	0,37	0,48	0,44	0,39	0,35	0,32	0,3	0,29	0,29	0,29	0,3	0,32	0,35	0,39	0,44	0,48	0,37	
-0,32	0,31	0,47	0,44	0,39	0,35	0,32	0,3	0,29	0,28	0,29	0,3	0,32	0,35	0,39	0,44	0,47	0,31	
-0,33	0,24	0,45	0,43	0,38	0,35	0,32	0,3	0,29	0,28	0,29	0,3	0,32	0,35	0,38	0,43	0,45	0,24	
-0,34	0,11	0,43	0,42	0,38	0,34	0,31	0,3	0,29	0,28	0,29	0,3	0,31	0,34	0,38	0,42	0,43	0,11	
-0,35		0,42	0,42	0,38	0,34	0,31	0,29	0,28	0,28	0,28	0,29	0,31	0,34	0,38	0,42	0,42		
-0,36		0,39	0,41	0,37	0,34	0,31	0,29	0,28	0,28	0,28	0,29	0,31	0,34	0,37	0,41	0,39		
-0,37		0,37	0,4	0,37	0,34	0,31	0,29	0,28	0,28	0,28	0,29	0,31	0,34	0,37	0,4	0,37		
-0,38		0,35	0,39	0,37	0,33	0,31	0,29	0,28	0,28	0,28	0,29	0,31	0,33	0,37	0,39	0,35		
-0,39		0,32	0,38	0,36	0,33	0,31	0,29	0,28	0,28	0,28	0,29	0,31	0,33	0,36	0,38	0,32		
-0,4		0,29	0,37	0,36	0,33	0,3	0,29	0,28	0,27	0,28	0,29	0,3	0,33	0,36	0,37	0,29		
-0,41		0,26	0,37	0,35	0,32	0,3	0,29	0,28	0,27	0,28	0,29	0,3	0,32	0,35	0,37	0,26		
-0,42		0,22	0,35	0,35	0,32	0,3	0,28	0,28	0,27	0,28	0,28	0,3	0,32	0,35	0,35	0,22		

	c1	c2	c3	c4	c5	c6	c7	c8	c9	c10	c11	c12	c13	c14	c15	c16	c17	c18
-0,43		0,16	0,34	0,34	0,32	0,3	0,28	0,27	0,27	0,27	0,28	0,3	0,32	0,34	0,34	0,16		
-0,44		0,08	0,33	0,34	0,32	0,3	0,28	0,27	0,27	0,27	0,28	0,3	0,32	0,34	0,33	0,08		
-0,45			0,32	0,33	0,31	0,29	0,28	0,27	0,27	0,27	0,28	0,29	0,31	0,33	0,32			
-0,46			0,31	0,32	0,31	0,29	0,28	0,27	0,27	0,27	0,28	0,29	0,31	0,32	0,31			
-0,47			0,29	0,32	0,31	0,29	0,28	0,27	0,26	0,27	0,28	0,29	0,31	0,32	0,29			
-0,48			0,28	0,31	0,3	0,29	0,27	0,27	0,26	0,27	0,27	0,29	0,3	0,31	0,28			
-0,49			0,26	0,31	0,3	0,28	0,27	0,26	0,26	0,26	0,27	0,28	0,3	0,31	0,26			
-0,5			0,25	0,3	0,29	0,28	0,27	0,26	0,26	0,26	0,27	0,28	0,29	0,3	0,25			
-0,51			0,23	0,29	0,29	0,28	0,27	0,26	0,26	0,26	0,27	0,28	0,29	0,29	0,23			
-0,52			0,21	0,28	0,29	0,28	0,27	0,26	0,26	0,26	0,27	0,28	0,29	0,28	0,21			
-0,53			0,18	0,28	0,28	0,27	0,26	0,26	0,25	0,26	0,26	0,27	0,28	0,28	0,18			
-0,54			0,15	0,27	0,28	0,27	0,26	0,25	0,25	0,25	0,26	0,27	0,28	0,27	0,15			
-0,55			0,12	0,26	0,27	0,27	0,26	0,25	0,25	0,25	0,26	0,27	0,27	0,26	0,12			
-0,56			0,06	0,25	0,27	0,26	0,26	0,25	0,25	0,25	0,26	0,26	0,27	0,25	0,06			
-0,57				0,24	0,26	0,26	0,25	0,25	0,25	0,25	0,26	0,26	0,26	0,24				
-0,58				0,23	0,26	0,26	0,25	0,25	0,24	0,25	0,25	0,26	0,26	0,23				
-0,59				0,22	0,25	0,25	0,25	0,24	0,24	0,24	0,25	0,25	0,25	0,22				
-0,6				0,21	0,25	0,25	0,25	0,24	0,24	0,24	0,25	0,25	0,25	0,21				
-0,61				0,2	0,24	0,25	0,24	0,24	0,24	0,24	0,24	0,25	0,24	0,2				
-0,62				0,18	0,24	0,24	0,24	0,24	0,24	0,24	0,24	0,24	0,24	0,18				
-0,63				0,17	0,23	0,24	0,24	0,23	0,23	0,23	0,24	0,24	0,23	0,17				
-0,64				0,15	0,22	0,23	0,23	0,23	0,23	0,23	0,23	0,23	0,22	0,15				
-0,65				0,13	0,22	0,23	0,23	0,23	0,23	0,23	0,23	0,23	0,22	0,13				
-0,66				0,11	0,21	0,23	0,23	0,23	0,23	0,23	0,23	0,23	0,21	0,11				
-0,67				0,08	0,2	0,22	0,22	0,22	0,22	0,22	0,22	0,22	0,2	0,08				
-0,68				0,04	0,19	0,22	0,22	0,22	0,22	0,22	0,22	0,22	0,19	0,04				
-0,69					0,18	0,21	0,22	0,22	0,22	0,22	0,22	0,21	0,18					
-0,7					0,18	0,21	0,21	0,21	0,21	0,21	0,21	0,21	0,18					
-0,71					0,17	0,21	0,21	0,21	0,21	0,21	0,21	0,2	0,17					
-0,72					0,16	0,2	0,21	0,21	0,21	0,21	0,21	0,2	0,16					
-0,73					0,15	0,19	0,2	0,2	0,21	0,2	0,2	0,19	0,15					
-0,74					0,13	0,2	0,2	0,2	0,2	0,2	0,2	0,18	0,13					
-0,75					0,12	0,18	0,19	0,2	0,2	0,2	0,19	0,18	0,12					
-0,76					0,1	0,17	0,19	0,19	0,19	0,19	0,19	0,17	0,1					
-0,77					0,08	0,16	0,18	0,19	0,19	0,19	0,18	0,16	0,08					
-0,78					0,06	0,16	0,18	0,19	0,19	0,19	0,18	0,16	0,06					
-0,79						0,15	0,17	0,18	0,18	0,18	0,17	0,15						
-0,8						0,14	0,17	0,18	0,18	0,18	0,17	0,14						
-0,81						0,13	0,16	0,17	0,18	0,17	0,16	0,13						
-0,82						0,12	0,16	0,17	0,17	0,17	0,16	0,12						
-0,83						0,11	0,15	0,16	0,17	0,16	0,15	0,11						
-0,84						0,1	0,15	0,16	0,16	0,16	0,15	0,1						
-0,85						0,09	0,14	0,15	0,16	0,15	0,14	0,09						
-0,86						0,07	0,13	0,15	0,15	0,15	0,13	0,07						
-0,87						0,05	0,12	0,14	0,15	0,14	0,12	0,05						
-0,88							0,12	0,14	0,14	0,14	0,12							
-0,89							0,11	0,13	0,14	0,13	0,11							
-0,9							0,1	0,12	0,13	0,12	0,1							
-0,91							0,09	0,12	0,12	0,12	0,09							
-0,92							0,07	0,11	0,12	0,11	0,07							
-0,93							0,06	0,1	0,11	0,1	0,06							
-0,94							0,03	0,09	0,1	0,09	0,03							
-0,95								0,08	0,09	0,08								
-0,96								0,07	0,08	0,07								
-0,97								0,05	0,07	0,05								
-0,98								0,03	0,06	0,03								
-0,99									0,04									
-1																		
$\sum_{\Delta\vec{h}_x} H_{\beta i/x}$	47,2	47,2	47,1	47,1	47,1	47,1	47,1	47,1	47,1	47,1	47,1	47,1	47,1	47,1	47,1	47,2	47,2	47
$n_{\beta,\Delta\vec{h}_x}$	69	89	113	137	157	175	189	197	199	197	189	175	157	137	113	89	69	59
$\bar{h}_{\beta/x}$	0,68	0,53	0,42	0,34	0,3	0,27	0,25	0,24	0,24	0,24	0,25	0,27	0,3	0,34	0,42	0,53	0,68	0,8
$d_{\beta/x}$	0,88	0,68	0,53	0,44	0,38	0,34	0,32	0,3	0,3	0,3	0,32	0,34	0,38	0,44	0,53	0,68	0,88	1
$\bar{h}_\beta/d_\beta$	0,78	0,78	0,78	0,78	0,79	0,79	0,79	0,79	0,79	0,79	0,79	0,79	0,79	0,78	0,78	0,78	0,78	0,8

$$\bar{\bar{h}}_{k\beta/x(\Delta\vec{h}_x)} = 0,39; \quad \widetilde{h}_{k/x,\Delta\vec{h}_x} = 0,34; \quad \bar{\bar{h}}_{k\beta/x(\Delta\vec{h}_x)}/\bar{d}_{k/x} = 0,786 \rightarrow \pi/4 = 0,785398$$

$$k = 0,1$$

Δh/a	0,17	0,35	0,52	0,7	0,87	1,05	1,22	1,4	1,57	1,75	1,92	2,09	2,27	2,44	2,62	2,79	2,97	3,14
1																		
0,99									0,01									
0,98								0,01	0,02	0,01								
0,97								0,02	0,02	0,02								
0,96								0,02	0,03	0,02								
0,95								0,03	0,03	0,03								
0,94								0,03	0,03	0,03								
0,93							0,02	0,03	0,04	0,03	0,02							
0,92							0,02	0,04	0,04	0,04	0,02							
0,91							0,03	0,04	0,04	0,04	0,03							
0,9							0,03	0,04	0,04	0,04	0,03							
0,89							0,03	0,04	0,05	0,04	0,03							
0,88							0,04	0,05	0,05	0,05	0,04							
0,87							0,04	0,05	0,05	0,05	0,04							
0,86						0,02	0,04	0,05	0,05	0,05	0,04	0,02						
0,85						0,02	0,05	0,05	0,05	0,05	0,05	0,02						
0,84						0,03	0,05	0,05	0,05	0,05	0,05	0,03						
0,83						0,03	0,05	0,05	0,06	0,05	0,05	0,03						
0,82						0,04	0,05	0,06	0,06	0,06	0,05	0,04						
0,81						0,04	0,05	0,06	0,06	0,06	0,05	0,04						
0,8						0,04	0,06	0,06	0,06	0,06	0,06	0,04						
0,79						0,05	0,06	0,06	0,06	0,06	0,06	0,05						
0,78						0,05	0,06	0,06	0,06	0,06	0,06	0,05						
0,77						0,05	0,06	0,06	0,06	0,06	0,06	0,05						
0,76					0,02	0,06	0,06	0,06	0,06	0,06	0,06	0,06	0,02					
0,75					0,03	0,06	0,06	0,07	0,07	0,07	0,06	0,06	0,03					
0,74					0,04	0,06	0,07	0,07	0,07	0,07	0,07	0,06	0,04					
0,73					0,04	0,06	0,07	0,07	0,07	0,07	0,07	0,06	0,04					
0,72					0,05	0,06	0,07	0,07	0,07	0,07	0,07	0,06	0,05					
0,71					0,05	0,07	0,07	0,07	0,07	0,07	0,07	0,07	0,05					
0,7					0,05	0,07	0,07	0,07	0,07	0,07	0,07	0,07	0,05					
0,69					0,06	0,07	0,07	0,07	0,07	0,07	0,07	0,07	0,06					
0,68					0,06	0,07	0,07	0,07	0,07	0,07	0,07	0,07	0,06					
0,67					0,06	0,07	0,07	0,07	0,07	0,07	0,07	0,07	0,06					
0,66					0,07	0,07	0,08	0,08	0,08	0,08	0,08	0,07	0,07					
0,65					0,07	0,08	0,08	0,08	0,08	0,08	0,08	0,08	0,07					
0,64				0,02	0,07	0,08	0,08	0,08	0,08	0,08	0,08	0,08	0,07	0,02				
0,63				0,04	0,07	0,08	0,08	0,08	0,08	0,08	0,08	0,08	0,07	0,04				
0,62				0,04	0,08	0,08	0,08	0,08	0,08	0,08	0,08	0,08	0,08	0,04				
0,61				0,05	0,08	0,08	0,08	0,08	0,08	0,08	0,08	0,08	0,08	0,05				
0,6				0,06	0,08	0,08	0,08	0,08	0,08	0,08	0,08	0,08	0,08	0,06				
0,59				0,06	0,08	0,08	0,08	0,08	0,08	0,08	0,08	0,08	0,08	0,06				
0,58				0,07	0,09	0,09	0,08	0,08	0,08	0,08	0,08	0,09	0,09	0,07				
0,57				0,07	0,09	0,09	0,08	0,08	0,08	0,08	0,08	0,09	0,09	0,07				
0,56				0,08	0,09	0,09	0,09	0,08	0,08	0,08	0,09	0,09	0,09	0,08				
0,55				0,08	0,09	0,09	0,09	0,08	0,08	0,08	0,09	0,09	0,09	0,08				
0,54				0,09	0,09	0,09	0,09	0,08	0,08	0,08	0,09	0,09	0,09	0,09				
0,53				0,09	0,09	0,09	0,09	0,08	0,09	0,09	0,09	0,09	0,09	0,09				
0,52				0,09	0,1	0,09	0,09	0,09	0,09	0,09	0,09	0,09	0,1	0,09				
0,51				0,1	0,1	0,09	0,09	0,09	0,09	0,09	0,09	0,09	0,1	0,1				
0,5			0,03	0,1	0,1	0,09	0,09	0,09	0,09	0,09	0,09	0,1	0,1	0,1	0,03			
0,49			0,05	0,1	0,1	0,1	0,09	0,09	0,09	0,09	0,09	0,1	0,1	0,1	0,05			
0,48			0,06	0,1	0,1	0,1	0,09	0,09	0,09	0,09	0,09	0,1	0,1	0,1	0,06			
0,47			0,07	0,11	0,1	0,1	0,09	0,09	0,09	0,09	0,09	0,1	0,1	0,11	0,07			
0,46			0,08	0,11	0,1	0,1	0,09	0,09	0,09	0,09	0,09	0,1	0,1	0,11	0,08			
0,45			0,09	0,11	0,11	0,1	0,09	0,09	0,09	0,09	0,09	0,1	0,11	0,11	0,09			
0,44			0,1	0,11	0,11	0,1	0,09	0,09	0,09	0,09	0,09	0,1	0,11	0,11	0,1			
0,43			0,1	0,12	0,11	0,1	0,09	0,09	0,09	0,09	0,09	0,1	0,11	0,12	0,1			
0,42			0,11	0,12	0,11	0,1	0,1	0,09	0,09	0,09	0,1	0,1	0,11	0,12	0,11			
0,41			0,12	0,12	0,11	0,1	0,1	0,09	0,09	0,09	0,1	0,1	0,11	0,12	0,12			
0,4			0,12	0,12	0,11	0,1	0,1	0,09	0,09	0,09	0,1	0,1	0,11	0,12	0,12			
0,39			0,13	0,12	0,11	0,1	0,1	0,09	0,09	0,09	0,1	0,1	0,11	0,12	0,13			
0,38			0,13	0,13	0,11	0,1	0,1	0,09	0,09	0,09	0,1	0,1	0,11	0,13	0,13			
0,37			0,13	0,13	0,11	0,1	0,1	0,09	0,09	0,09	0,1	0,1	0,11	0,13	0,13			
0,36			0,14	0,13	0,11	0,1	0,1	0,09	0,09	0,09	0,1	0,1	0,11	0,13	0,14			
0,35		0,05	0,14	0,13	0,12	0,11	0,1	0,09	0,09	0,09	0,1	0,11	0,12	0,13	0,14	0,05		
0,34		0,08	0,15	0,13	0,12	0,11	0,1	0,1	0,09	0,1	0,1	0,11	0,12	0,13	0,15	0,08		

x	L																R	R2
0,33		0,1	0,15	0,13	0,12	0,11	0,1	0,1	0,09	0,1	0,1	0,11	0,12	0,13	0,15	0,1		
0,32		0,12	0,15	0,13	0,12	0,11	0,1	0,1	0,09	0,1	0,1	0,11	0,12	0,13	0,15	0,12		
0,31		0,14	0,16	0,14	0,12	0,11	0,1	0,1	0,1	0,1	0,1	0,11	0,12	0,14	0,16	0,14		
0,3		0,15	0,16	0,14	0,12	0,11	0,1	0,1	0,1	0,1	0,1	0,11	0,12	0,14	0,16	0,15		
0,29		0,16	0,16	0,14	0,12	0,11	0,1	0,1	0,1	0,1	0,1	0,11	0,12	0,14	0,16	0,16		
0,28		0,17	0,16	0,14	0,12	0,11	0,1	0,1	0,1	0,1	0,1	0,11	0,12	0,14	0,16	0,17		
0,27		0,18	0,17	0,14	0,12	0,11	0,1	0,1	0,1	0,1	0,1	0,11	0,12	0,14	0,17	0,18		
0,26		0,19	0,17	0,14	0,12	0,11	0,1	0,1	0,1	0,1	0,1	0,11	0,12	0,14	0,17	0,19		
0,25		0,2	0,17	0,14	0,12	0,11	0,1	0,1	0,1	0,1	0,1	0,11	0,12	0,14	0,17	0,2		
0,24		0,21	0,17	0,14	0,12	0,11	0,1	0,1	0,1	0,1	0,1	0,11	0,12	0,14	0,17	0,21		
0,23		0,21	0,18	0,14	0,12	0,11	0,1	0,1	0,1	0,1	0,1	0,11	0,12	0,14	0,18	0,21		
0,22		0,22	0,18	0,15	0,12	0,11	0,1	0,1	0,1	0,1	0,1	0,11	0,12	0,15	0,18	0,22		
0,21		0,23	0,18	0,15	0,13	0,11	0,1	0,1	0,1	0,1	0,1	0,11	0,13	0,15	0,18	0,23		
0,2		0,23	0,18	0,15	0,13	0,11	0,1	0,1	0,1	0,1	0,1	0,11	0,13	0,15	0,18	0,23		
0,19	0,15	0,24	0,18	0,15	0,13	0,11	0,1	0,1	0,1	0,1	0,1	0,11	0,13	0,15	0,18	0,24	0,15	
0,18	0,22	0,24	0,18	0,15	0,13	0,11	0,1	0,1	0,1	0,1	0,1	0,11	0,13	0,15	0,18	0,24	0,22	
0,17	0,26	0,25	0,19	0,15	0,13	0,11	0,1	0,1	0,1	0,1	0,1	0,11	0,13	0,15	0,19	0,25	0,26	
0,16	0,3	0,25	0,19	0,15	0,13	0,11	0,1	0,1	0,1	0,1	0,1	0,11	0,13	0,15	0,19	0,25	0,3	
0,15	0,33	0,26	0,19	0,15	0,13	0,11	0,1	0,1	0,1	0,1	0,1	0,11	0,13	0,15	0,19	0,26	0,33	
0,14	0,36	0,26	0,19	0,15	0,13	0,11	0,11	0,1	0,1	0,1	0,11	0,11	0,13	0,15	0,19	0,26	0,36	
0,13	0,38	0,26	0,19	0,15	0,13	0,11	0,11	0,1	0,1	0,1	0,11	0,11	0,13	0,15	0,19	0,26	0,38	
0,12	0,4	0,27	0,19	0,15	0,13	0,11	0,11	0,1	0,1	0,1	0,11	0,11	0,13	0,15	0,19	0,27	0,4	
0,11	0,42	0,27	0,19	0,15	0,13	0,11	0,11	0,1	0,1	0,1	0,11	0,11	0,13	0,15	0,19	0,27	0,42	
0,1	0,43	0,27	0,19	0,15	0,13	0,11	0,11	0,1	0,1	0,1	0,11	0,11	0,13	0,15	0,19	0,27	0,43	
0,09	0,45	0,27	0,19	0,15	0,13	0,11	0,11	0,1	0,1	0,1	0,11	0,11	0,13	0,15	0,19	0,27	0,45	0,44
0,08	0,46	0,27	0,19	0,15	0,13	0,11	0,11	0,1	0,1	0,1	0,11	0,11	0,13	0,15	0,19	0,27	0,46	0,6
0,07	0,47	0,28	0,2	0,15	0,13	0,11	0,11	0,1	0,1	0,1	0,11	0,11	0,13	0,15	0,2	0,28	0,47	0,71
0,06	0,48	0,28	0,2	0,15	0,13	0,12	0,11	0,1	0,1	0,1	0,11	0,12	0,13	0,15	0,2	0,28	0,48	0,8
0,05	0,48	0,28	0,2	0,15	0,13	0,12	0,11	0,1	0,1	0,1	0,11	0,12	0,13	0,15	0,2	0,28	0,48	0,87
0,04	0,49	0,28	0,2	0,15	0,13	0,12	0,11	0,1	0,1	0,1	0,11	0,12	0,13	0,15	0,2	0,28	0,49	0,92
0,03	0,5	0,28	0,2	0,15	0,13	0,12	0,11	0,1	0,1	0,1	0,11	0,12	0,13	0,15	0,2	0,28	0,5	0,95
0,02	0,5	0,28	0,2	0,15	0,13	0,12	0,11	0,1	0,1	0,1	0,11	0,12	0,13	0,15	0,2	0,28	0,5	0,98
0,01	0,5	0,28	0,2	0,15	0,13	0,12	0,11	0,1	0,1	0,1	0,11	0,12	0,13	0,15	0,2	0,28	0,5	0,99
-0	0,5	0,28	0,2	0,15	0,13	0,12	0,11	0,1	0,1	0,1	0,11	0,12	0,13	0,15	0,2	0,28	0,5	1
-0,01	0,5	0,28	0,2	0,15	0,13	0,12	0,11	0,1	0,1	0,1	0,11	0,12	0,13	0,15	0,2	0,28	0,5	0,99
-0,02	0,5	0,28	0,2	0,15	0,13	0,12	0,11	0,1	0,1	0,1	0,11	0,12	0,13	0,15	0,2	0,28	0,5	0,98
-0,03	0,5	0,28	0,2	0,15	0,13	0,12	0,11	0,1	0,1	0,1	0,11	0,12	0,13	0,15	0,2	0,28	0,5	0,95
-0,04	0,49	0,28	0,2	0,15	0,13	0,12	0,11	0,1	0,1	0,1	0,11	0,12	0,13	0,15	0,2	0,28	0,49	0,92
-0,05	0,48	0,28	0,2	0,15	0,13	0,12	0,11	0,1	0,1	0,1	0,11	0,12	0,13	0,15	0,2	0,28	0,48	0,87
-0,06	0,48	0,28	0,2	0,15	0,13	0,12	0,11	0,1	0,1	0,1	0,11	0,12	0,13	0,15	0,2	0,28	0,48	0,8
-0,07	0,47	0,28	0,2	0,15	0,13	0,11	0,11	0,1	0,1	0,1	0,11	0,11	0,13	0,15	0,2	0,28	0,47	0,71
-0,08	0,46	0,27	0,19	0,15	0,13	0,11	0,11	0,1	0,1	0,1	0,11	0,11	0,13	0,15	0,19	0,27	0,46	0,6
-0,09	0,45	0,27	0,19	0,15	0,13	0,11	0,11	0,1	0,1	0,1	0,11	0,11	0,13	0,15	0,19	0,27	0,45	0,44
-0,1	0,43	0,27	0,19	0,15	0,13	0,11	0,11	0,1	0,1	0,1	0,11	0,11	0,13	0,15	0,19	0,27	0,43	
-0,11	0,42	0,27	0,19	0,15	0,13	0,11	0,11	0,1	0,1	0,1	0,11	0,11	0,13	0,15	0,19	0,27	0,42	
-0,12	0,4	0,27	0,19	0,15	0,13	0,11	0,11	0,1	0,1	0,1	0,11	0,11	0,13	0,15	0,19	0,27	0,4	
-0,13	0,38	0,26	0,19	0,15	0,13	0,11	0,11	0,1	0,1	0,1	0,11	0,11	0,13	0,15	0,19	0,26	0,38	
-0,14	0,36	0,26	0,19	0,15	0,13	0,11	0,11	0,1	0,1	0,1	0,11	0,11	0,13	0,15	0,19	0,26	0,36	
-0,15	0,33	0,26	0,19	0,15	0,13	0,11	0,1	0,1	0,1	0,1	0,1	0,11	0,13	0,15	0,19	0,26	0,33	
-0,16	0,3	0,25	0,19	0,15	0,13	0,11	0,1	0,1	0,1	0,1	0,1	0,11	0,13	0,15	0,19	0,25	0,3	
-0,17	0,26	0,25	0,19	0,15	0,13	0,11	0,1	0,1	0,1	0,1	0,1	0,11	0,13	0,15	0,19	0,25	0,26	
-0,18	0,22	0,24	0,18	0,15	0,13	0,11	0,1	0,1	0,1	0,1	0,1	0,11	0,13	0,15	0,18	0,24	0,22	
-0,19	0,15	0,24	0,18	0,15	0,13	0,11	0,1	0,1	0,1	0,1	0,1	0,11	0,13	0,15	0,18	0,24	0,15	
-0,2		0,23	0,18	0,15	0,13	0,11	0,1	0,1	0,1	0,1	0,1	0,11	0,13	0,15	0,18	0,23		
-0,21		0,23	0,18	0,15	0,13	0,11	0,1	0,1	0,1	0,1	0,1	0,11	0,13	0,15	0,18	0,23		
-0,22		0,22	0,18	0,15	0,12	0,11	0,1	0,1	0,1	0,1	0,1	0,11	0,12	0,15	0,18	0,22		
-0,23		0,21	0,18	0,14	0,12	0,11	0,1	0,1	0,1	0,1	0,1	0,11	0,12	0,14	0,18	0,21		
-0,24		0,21	0,17	0,14	0,12	0,11	0,1	0,1	0,1	0,1	0,1	0,11	0,12	0,14	0,17	0,21		
-0,25		0,2	0,17	0,14	0,12	0,11	0,1	0,1	0,1	0,1	0,1	0,11	0,12	0,14	0,17	0,2		
-0,26		0,19	0,17	0,14	0,12	0,11	0,1	0,1	0,1	0,1	0,1	0,11	0,12	0,14	0,17	0,19		
-0,27		0,18	0,17	0,14	0,12	0,11	0,1	0,1	0,1	0,1	0,1	0,11	0,12	0,14	0,17	0,18		
-0,28		0,17	0,16	0,14	0,12	0,11	0,1	0,1	0,1	0,1	0,1	0,11	0,12	0,14	0,16	0,17		
-0,29		0,16	0,16	0,14	0,12	0,11	0,1	0,1	0,1	0,1	0,1	0,11	0,12	0,14	0,16	0,16		
-0,3		0,15	0,16	0,14	0,12	0,11	0,1	0,1	0,1	0,1	0,1	0,11	0,12	0,14	0,16	0,15		
-0,31		0,14	0,16	0,14	0,12	0,11	0,1	0,1	0,1	0,1	0,1	0,11	0,12	0,14	0,16	0,14		
-0,32		0,12	0,15	0,13	0,12	0,11	0,1	0,1	0,09	0,1	0,1	0,11	0,12	0,13	0,15	0,12		
-0,33		0,1	0,15	0,13	0,12	0,11	0,1	0,1	0,09	0,1	0,1	0,11	0,12	0,13	0,15	0,1		
-0,34		0,08	0,15	0,13	0,12	0,11	0,1	0,1	0,09	0,1	0,1	0,11	0,12	0,13	0,15	0,08		
-0,35	0,05	0,14	0,13	0,12	0,11	0,1	0,09	0,09	0,09	0,1	0,11	0,12	0,13	0,14			0,05	
-0,36		0,14	0,13	0,11	0,1	0,1	0,09	0,09	0,09	0,1	0,1	0,11	0,13	0,14				
-0,37		0,13	0,13	0,11	0,1	0,1	0,09	0,09	0,09	0,1	0,1	0,11	0,13	0,13				
-0,38		0,13	0,13	0,11	0,1	0,1	0,09	0,09	0,09	0,1	0,1	0,11	0,13	0,13				
-0,39		0,13	0,12	0,11	0,1	0,1	0,09	0,09	0,09	0,1	0,1	0,11	0,12	0,13				
-0,4		0,12	0,12	0,11	0,1	0,1	0,09	0,09	0,09	0,1	0,1	0,11	0,12	0,12				
-0,41		0,12	0,12	0,11	0,1	0,1	0,09	0,09	0,09	0,1	0,1	0,11	0,12	0,12				
-0,42		0,11	0,12	0,11	0,1	0,1	0,09	0,09	0,09	0,1	0,1	0,11	0,12	0,11				

-0,43			0,1	0,12	0,11	0,1	0,09	0,09	0,09	0,09	0,09	0,1	0,11	0,12	0,1			
-0,44			0,1	0,11	0,11	0,1	0,09	0,09	0,09	0,09	0,09	0,1	0,11	0,11	0,1			
-0,45			0,09	0,11	0,11	0,1	0,09	0,09	0,09	0,09	0,09	0,1	0,11	0,11	0,09			
-0,46			0,08	0,11	0,1	0,1	0,09	0,09	0,09	0,09	0,09	0,1	0,1	0,11	0,08			
-0,47			0,07	0,11	0,1	0,1	0,09	0,09	0,09	0,09	0,09	0,1	0,1	0,11	0,07			
-0,48			0,06	0,1	0,1	0,1	0,09	0,09	0,09	0,09	0,09	0,1	0,1	0,1	0,06			
-0,49			0,05	0,1	0,1	0,1	0,09	0,09	0,09	0,09	0,09	0,1	0,1	0,1	0,05			
-0,5			0,03	0,1	0,1	0,09	0,09	0,09	0,09	0,09	0,09	0,09	0,1	0,1	0,03			
-0,51				0,1	0,1	0,09	0,09	0,09	0,09	0,09	0,09	0,09	0,1	0,1				
-0,52				0,09	0,1	0,09	0,09	0,09	0,09	0,09	0,09	0,09	0,1	0,09				
-0,53				0,09	0,09	0,09	0,09	0,09	0,08	0,09	0,09	0,09	0,09	0,09				
-0,54				0,09	0,09	0,09	0,09	0,08	0,08	0,08	0,09	0,09	0,09	0,09				
-0,55				0,08	0,09	0,09	0,09	0,08	0,08	0,08	0,09	0,09	0,09	0,08				
-0,56				0,08	0,09	0,09	0,09	0,08	0,08	0,08	0,09	0,09	0,09	0,08				
-0,57				0,07	0,09	0,09	0,08	0,08	0,08	0,08	0,08	0,09	0,09	0,07				
-0,58				0,07	0,09	0,09	0,08	0,08	0,08	0,08	0,08	0,09	0,09	0,07				
-0,59				0,06	0,08	0,08	0,08	0,08	0,08	0,08	0,08	0,08	0,08	0,06				
-0,6				0,06	0,08	0,08	0,08	0,08	0,08	0,08	0,08	0,08	0,08	0,06				
-0,61				0,05	0,08	0,08	0,08	0,08	0,08	0,08	0,08	0,08	0,08	0,05				
-0,62				0,04	0,08	0,08	0,08	0,08	0,08	0,08	0,08	0,08	0,08	0,04				
-0,63				0,04	0,07	0,08	0,08	0,08	0,08	0,08	0,08	0,08	0,07	0,04				
-0,64				0,02	0,07	0,08	0,08	0,08	0,08	0,08	0,08	0,08	0,07	0,02				
-0,65					0,07	0,08	0,08	0,08	0,08	0,08	0,08	0,08	0,07					
-0,66					0,07	0,07	0,08	0,08	0,08	0,08	0,08	0,07	0,07					
-0,67					0,06	0,07	0,07	0,07	0,07	0,07	0,07	0,07	0,06					
-0,68					0,06	0,07	0,07	0,07	0,07	0,07	0,07	0,07	0,06					
-0,69					0,06	0,07	0,07	0,07	0,07	0,07	0,07	0,07	0,06					
-0,7					0,05	0,07	0,07	0,07	0,07	0,07	0,07	0,07	0,05					
-0,71					0,05	0,07	0,07	0,07	0,07	0,07	0,07	0,07	0,05					
-0,72					0,05	0,06	0,07	0,07	0,07	0,07	0,07	0,06	0,05					
-0,73					0,04	0,06	0,07	0,07	0,07	0,07	0,07	0,06	0,04					
-0,74					0,04	0,06	0,07	0,07	0,07	0,07	0,07	0,06	0,04					
-0,75					0,03	0,06	0,06	0,07	0,07	0,07	0,06	0,06	0,03					
-0,76					0,02	0,06	0,06	0,06	0,06	0,06	0,06	0,06	0,02					
-0,77						0,05	0,06	0,06	0,06	0,06	0,06	0,05						
-0,78						0,05	0,06	0,06	0,06	0,06	0,06	0,05						
-0,79						0,05	0,06	0,06	0,06	0,06	0,06	0,05						
-0,8						0,04	0,06	0,06	0,06	0,06	0,06	0,04						
-0,81						0,04	0,05	0,06	0,06	0,06	0,05	0,04						
-0,82						0,04	0,05	0,06	0,06	0,06	0,05	0,04						
-0,83						0,03	0,05	0,05	0,06	0,05	0,05	0,03						
-0,84						0,03	0,05	0,05	0,05	0,05	0,05	0,03						
-0,85						0,02	0,05	0,05	0,05	0,05	0,05	0,02						
-0,86						0,02	0,04	0,05	0,05	0,05	0,04	0,02						
-0,87							0,04	0,05	0,05	0,05	0,04							
-0,88							0,04	0,05	0,05	0,05	0,04							
-0,89							0,03	0,04	0,05	0,04	0,03							
-0,9							0,03	0,04	0,04	0,04	0,03							
-0,91							0,03	0,04	0,04	0,04	0,03							
-0,92							0,02	0,04	0,04	0,04	0,02							
-0,93							0,02	0,03	0,04	0,03	0,02							
-0,94								0,03	0,03	0,03								
-0,95								0,03	0,03	0,03								
-0,96								0,02	0,03	0,02								
-0,97								0,02	0,02	0,02								
-0,98								0,01	0,02	0,01								
-0,99									0,01									
-1																		
$\sum_{\Delta\vec{h}_x} H_{\beta i/x}$	15,7	15,7	15,7	15,7	15,7	15,7	15,7	15,7	15,7	15,7	15,7	15,7	15,7	15,7	15,7	15,7	15,7	15,5
$n_{\beta,\Delta\vec{h}_x}$	39	71	101	129	153	173	187	197	199	197	187	173	153	129	101	71	39	19
$\bar{h}_{\beta/x}$	0,4	0,22	0,16	0,12	0,1	0,09	0,08	0,08	0,08	0,08	0,08	0,09	0,1	0,12	0,16	0,22	0,4	0,82
$d_{\beta/x}$	0,5	0,28	0,2	0,15	0,13	0,12	0,11	0,1	0,1	0,1	0,11	0,12	0,13	0,15	0,2	0,28	0,5	1
$\bar{h}_\beta/d_\beta$	0,8	0,79	0,79	0,79	0,79	0,79	0,79	0,79	0,79	0,79	0,79	0,79	0,79	0,79	0,79	0,79	0,8	0,82

$$\bar{\bar{h}}_{k\beta/x(\Delta\vec{h}_x)} = 0,19; \quad \tilde{h}_{k/x,\Delta\vec{h}_x} = 0,12; \quad \bar{\bar{h}}_{k\beta/x(\Delta\vec{h}_x)}/\bar{d}_{k/x} = 0,791 \rightarrow \pi/4 = 0,785398$$

$$k = 0,01$$

k 0,01 Δh/a	0,17	0,35	0,52	0,7	0,87	1,05	1,22	1,4	1,57	1,75	1,92	2,09	2,27	2,44	2,62	2,79	2,97	3,14
1																		
0,99																		
0,98																		
0,97																		
0,96																		
0,95																		
0,94																		
0,93																		
0,92																		
0,91																		
0,9																		
0,89																		
0,88																		
0,87																		
0,86									0,01									
0,85								0,01	0,01	0,01								
0,84								0,01	0,01	0,01								
0,83								0,01	0,01	0,01								
0,82							0,01	0,01	0,01	0,01	0,01							
0,81							0,01	0,01	0,01	0,01	0,01							
0,8							0,01	0,01	0,01	0,01	0,01							
0,79							0,01	0,01	0,01	0,01	0,01							
0,78						0,01	0,01	0,01	0,01	0,01	0,01	0,01						
0,77						0,01	0,01	0,01	0,01	0,01	0,01	0,01						
0,76						0,01	0,01	0,01	0,01	0,01	0,01	0,01						
0,75						0,01	0,01	0,01	0,01	0,01	0,01	0,01						
0,74						0,01	0,01	0,01	0,01	0,01	0,01	0,01						
0,73						0,01	0,01	0,01	0,01	0,01	0,01	0,01						
0,72						0,01	0,01	0,01	0,01	0,01	0,01	0,01						
0,71						0,01	0,01	0,01	0,01	0,01	0,01	0,01						
0,7					0,01	0,01	0,01	0,01	0,01	0,01	0,01	0,01	0,01					
0,69					0,01	0,01	0,01	0,01	0,01	0,01	0,01	0,01	0,01					
0,68					0,01	0,01	0,01	0,01	0,01	0,01	0,01	0,01	0,01					
0,67					0,01	0,01	0,01	0,01	0,01	0,01	0,01	0,01	0,01					
0,66					0,01	0,01	0,01	0,01	0,01	0,01	0,01	0,01	0,01					
0,65					0,01	0,01	0,01	0,01	0,01	0,01	0,01	0,01	0,01					
0,64					0,01	0,01	0,01	0,01	0,01	0,01	0,01	0,01	0,01					
0,63					0,01	0,01	0,01	0,01	0,01	0,01	0,01	0,01	0,01					
0,62					0,01	0,01	0,01	0,01	0,01	0,01	0,01	0,01	0,01					
0,61					0,01	0,01	0,01	0,01	0,01	0,01	0,01	0,01	0,01					
0,6				0,01	0,01	0,01	0,01	0,01	0,01	0,01	0,01	0,01	0,01	0,01				
0,59				0,01	0,01	0,01	0,01	0,01	0,01	0,01	0,01	0,01	0,01	0,01				
0,58				0,01	0,01	0,01	0,01	0,01	0,01	0,01	0,01	0,01	0,01	0,01				
0,57				0,01	0,01	0,01	0,01	0,01	0,01	0,01	0,01	0,01	0,01	0,01				
0,56				0,01	0,01	0,01	0,01	0,01	0,01	0,01	0,01	0,01	0,01	0,01				
0,55				0,01	0,01	0,01	0,01	0,01	0,01	0,01	0,01	0,01	0,01	0,01				
0,54				0,01	0,01	0,01	0,01	0,01	0,01	0,01	0,01	0,01	0,01	0,01				
0,53				0,01	0,01	0,01	0,01	0,01	0,01	0,01	0,01	0,01	0,01	0,01				
0,52				0,01	0,01	0,01	0,01	0,01	0,01	0,01	0,01	0,01	0,01	0,01				
0,51				0,01	0,01	0,01	0,01	0,01	0,01	0,01	0,01	0,01	0,01	0,01				
0,5				0,01	0,01	0,01	0,01	0,01	0,01	0,01	0,01	0,01	0,01	0,01				
0,49				0,01	0,01	0,01	0,01	0,01	0,01	0,01	0,01	0,01	0,01	0,01				
0,48			0,01	0,01	0,01	0,01	0,01	0,01	0,01	0,01	0,01	0,01	0,01	0,01	0,01			
0,47			0,01	0,01	0,01	0,01	0,01	0,01	0,01	0,01	0,01	0,01	0,01	0,01	0,01			
0,46			0,01	0,01	0,01	0,01	0,01	0,01	0,01	0,01	0,01	0,01	0,01	0,01	0,01			
0,45			0,01	0,01	0,01	0,01	0,01	0,01	0,01	0,01	0,01	0,01	0,01	0,01	0,01			
0,44			0,01	0,01	0,01	0,01	0,01	0,01	0,01	0,01	0,01	0,01	0,01	0,01	0,01			
0,43			0,01	0,01	0,01	0,01	0,01	0,01	0,01	0,01	0,01	0,01	0,01	0,01	0,01			
0,42			0,01	0,01	0,01	0,01	0,01	0,01	0,01	0,01	0,01	0,01	0,01	0,01	0,01			
0,41			0,01	0,01	0,01	0,01	0,01	0,01	0,01	0,01	0,01	0,01	0,01	0,01	0,01			
0,4			0,01	0,01	0,01	0,01	0,01	0,01	0,01	0,01	0,01	0,01	0,01	0,01	0,01			
0,39			0,01	0,01	0,01	0,01	0,01	0,01	0,01	0,01	0,01	0,01	0,01	0,01	0,01			
0,38			0,01	0,01	0,01	0,01	0,01	0,01	0,01	0,01	0,01	0,01	0,01	0,01	0,01			
0,37			0,01	0,01	0,01	0,01	0,01	0,01	0,01	0,01	0,01	0,01	0,01	0,01	0,01			
0,36			0,01	0,01	0,01	0,01	0,01	0,01	0,01	0,01	0,01	0,01	0,01	0,01	0,01			
0,35			0,01	0,01	0,01	0,01	0,01	0,01	0,01	0,01	0,01	0,01	0,01	0,01	0,01			
0,34			0,01	0,01	0,01	0,01	0,01	0,01	0,01	0,01	0,01	0,01	0,01	0,01	0,01			

0,33		0,01	0,02	0,01	0,01	0,01	0,01	0,01	0,01	0,01	0,01	0,01	0,01	0,01	0,02	0,01	
0,32		0,01	0,02	0,01	0,01	0,01	0,01	0,01	0,01	0,01	0,01	0,01	0,01	0,01	0,02	0,01	
0,31		0,01	0,02	0,01	0,01	0,01	0,01	0,01	0,01	0,01	0,01	0,01	0,01	0,01	0,02	0,01	
0,3		0,01	0,02	0,01	0,01	0,01	0,01	0,01	0,01	0,01	0,01	0,01	0,01	0,01	0,02	0,01	
0,29		0,02	0,02	0,01	0,01	0,01	0,01	0,01	0,01	0,01	0,01	0,01	0,01	0,01	0,02	0,02	
0,28		0,02	0,02	0,01	0,01	0,01	0,01	0,01	0,01	0,01	0,01	0,01	0,01	0,01	0,02	0,02	
0,27		0,02	0,02	0,01	0,01	0,01	0,01	0,01	0,01	0,01	0,01	0,01	0,01	0,01	0,02	0,02	
0,26		0,02	0,02	0,01	0,01	0,01	0,01	0,01	0,01	0,01	0,01	0,01	0,01	0,01	0,02	0,02	
0,25		0,02	0,02	0,01	0,01	0,01	0,01	0,01	0,01	0,01	0,01	0,01	0,01	0,01	0,02	0,02	
0,24		0,02	0,02	0,01	0,01	0,01	0,01	0,01	0,01	0,01	0,01	0,01	0,01	0,01	0,02	0,02	
0,23		0,02	0,02	0,01	0,01	0,01	0,01	0,01	0,01	0,01	0,01	0,01	0,01	0,01	0,02	0,02	
0,22		0,02	0,02	0,01	0,01	0,01	0,01	0,01	0,01	0,01	0,01	0,01	0,01	0,01	0,02	0,02	
0,21		0,02	0,02	0,01	0,01	0,01	0,01	0,01	0,01	0,01	0,01	0,01	0,01	0,01	0,02	0,02	
0,2		0,02	0,02	0,01	0,01	0,01	0,01	0,01	0,01	0,01	0,01	0,01	0,01	0,01	0,02	0,02	
0,19		0,02	0,02	0,01	0,01	0,01	0,01	0,01	0,01	0,01	0,01	0,01	0,01	0,01	0,02	0,02	
0,18		0,02	0,02	0,01	0,01	0,01	0,01	0,01	0,01	0,01	0,01	0,01	0,01	0,01	0,02	0,02	
0,17	0,01	0,03	0,02	0,02	0,01	0,01	0,01	0,01	0,01	0,01	0,01	0,01	0,01	0,02	0,02	0,03	0,01
0,16	0,02	0,03	0,02	0,02	0,01	0,01	0,01	0,01	0,01	0,01	0,01	0,01	0,01	0,02	0,02	0,03	0,02
0,15	0,03	0,03	0,02	0,02	0,01	0,01	0,01	0,01	0,01	0,01	0,01	0,01	0,01	0,02	0,02	0,03	0,03
0,14	0,03	0,03	0,02	0,02	0,01	0,01	0,01	0,01	0,01	0,01	0,01	0,01	0,01	0,02	0,02	0,03	0,03
0,13	0,04	0,03	0,02	0,02	0,01	0,01	0,01	0,01	0,01	0,01	0,01	0,01	0,01	0,02	0,02	0,03	0,04
0,12	0,04	0,03	0,02	0,02	0,01	0,01	0,01	0,01	0,01	0,01	0,01	0,01	0,01	0,02	0,02	0,03	0,04
0,11	0,04	0,03	0,02	0,02	0,01	0,01	0,01	0,01	0,01	0,01	0,01	0,01	0,01	0,02	0,02	0,03	0,04
0,1	0,05	0,03	0,02	0,02	0,01	0,01	0,01	0,01	0,01	0,01	0,01	0,01	0,01	0,02	0,02	0,03	0,05
0,09	0,05	0,03	0,02	0,02	0,01	0,01	0,01	0,01	0,01	0,01	0,01	0,01	0,01	0,02	0,02	0,03	0,05
0,08	0,05	0,03	0,02	0,02	0,01	0,01	0,01	0,01	0,01	0,01	0,01	0,01	0,01	0,02	0,02	0,03	0,05
0,07	0,05	0,03	0,02	0,02	0,01	0,01	0,01	0,01	0,01	0,01	0,01	0,01	0,01	0,02	0,02	0,03	0,05
0,06	0,05	0,03	0,02	0,02	0,01	0,01	0,01	0,01	0,01	0,01	0,01	0,01	0,01	0,02	0,02	0,03	0,05
0,05	0,06	0,03	0,02	0,02	0,01	0,01	0,01	0,01	0,01	0,01	0,01	0,01	0,01	0,02	0,02	0,03	0,06
0,04	0,06	0,03	0,02	0,02	0,01	0,01	0,01	0,01	0,01	0,01	0,01	0,01	0,01	0,02	0,02	0,03	0,06
0,03	0,06	0,03	0,02	0,02	0,01	0,01	0,01	0,01	0,01	0,01	0,01	0,01	0,01	0,02	0,02	0,03	0,06
0,02	0,06	0,03	0,02	0,02	0,01	0,01	0,01	0,01	0,01	0,01	0,01	0,01	0,01	0,02	0,02	0,03	0,06
0,01	0,06	0,03	0,02	0,02	0,01	0,01	0,01	0,01	0,01	0,01	0,01	0,01	0,01	0,02	0,02	0,03	0,06
-0	0,06	0,03	0,02	0,02	0,01	0,01	0,01	0,01	0,01	0,01	0,01	0,01	0,01	0,02	0,02	0,03	0,06	1
-0,01	0,06	0,03	0,02	0,02	0,01	0,01	0,01	0,01	0,01	0,01	0,01	0,01	0,01	0,02	0,02	0,03	0,06
-0,02	0,06	0,03	0,02	0,02	0,01	0,01	0,01	0,01	0,01	0,01	0,01	0,01	0,01	0,02	0,02	0,03	0,06
-0,03	0,06	0,03	0,02	0,02	0,01	0,01	0,01	0,01	0,01	0,01	0,01	0,01	0,01	0,02	0,02	0,03	0,06
-0,04	0,06	0,03	0,02	0,02	0,01	0,01	0,01	0,01	0,01	0,01	0,01	0,01	0,01	0,02	0,02	0,03	0,06
-0,05	0,06	0,03	0,02	0,02	0,01	0,01	0,01	0,01	0,01	0,01	0,01	0,01	0,01	0,02	0,02	0,03	0,06
-0,06	0,05	0,03	0,02	0,02	0,01	0,01	0,01	0,01	0,01	0,01	0,01	0,01	0,01	0,02	0,02	0,03	0,05
-0,07	0,05	0,03	0,02	0,02	0,01	0,01	0,01	0,01	0,01	0,01	0,01	0,01	0,01	0,02	0,02	0,03	0,05
-0,08	0,05	0,03	0,02	0,02	0,01	0,01	0,01	0,01	0,01	0,01	0,01	0,01	0,01	0,02	0,02	0,03	0,05
-0,09	0,05	0,03	0,02	0,02	0,01	0,01	0,01	0,01	0,01	0,01	0,01	0,01	0,01	0,02	0,02	0,03	0,05
-0,1	0,05	0,03	0,02	0,02	0,01	0,01	0,01	0,01	0,01	0,01	0,01	0,01	0,01	0,02	0,02	0,03	0,05
-0,11	0,04	0,03	0,02	0,02	0,01	0,01	0,01	0,01	0,01	0,01	0,01	0,01	0,01	0,02	0,02	0,03	0,04
-0,12	0,04	0,03	0,02	0,02	0,01	0,01	0,01	0,01	0,01	0,01	0,01	0,01	0,01	0,02	0,02	0,03	0,04
-0,13	0,04	0,03	0,02	0,02	0,01	0,01	0,01	0,01	0,01	0,01	0,01	0,01	0,01	0,02	0,02	0,03	0,04
-0,14	0,03	0,03	0,02	0,02	0,01	0,01	0,01	0,01	0,01	0,01	0,01	0,01	0,01	0,02	0,02	0,03	0,03
-0,15	0,03	0,03	0,02	0,02	0,01	0,01	0,01	0,01	0,01	0,01	0,01	0,01	0,01	0,02	0,02	0,03	0,03
-0,16	0,02	0,03	0,02	0,02	0,01	0,01	0,01	0,01	0,01	0,01	0,01	0,01	0,01	0,02	0,02	0,03	0,02
-0,17	0,01	0,03	0,02	0,02	0,01	0,01	0,01	0,01	0,01	0,01	0,01	0,01	0,01	0,02	0,02	0,03	0,01
-0,18		0,02	0,02	0,01	0,01	0,01	0,01	0,01	0,01	0,01	0,01	0,01	0,01	0,01	0,02	0,02	
-0,19		0,02	0,02	0,01	0,01	0,01	0,01	0,01	0,01	0,01	0,01	0,01	0,01	0,01	0,02	0,02	
-0,2		0,02	0,02	0,01	0,01	0,01	0,01	0,01	0,01	0,01	0,01	0,01	0,01	0,01	0,02	0,02	
-0,21		0,02	0,02	0,01	0,01	0,01	0,01	0,01	0,01	0,01	0,01	0,01	0,01	0,01	0,02	0,02	
-0,22		0,02	0,02	0,01	0,01	0,01	0,01	0,01	0,01	0,01	0,01	0,01	0,01	0,01	0,02	0,02	
-0,23		0,02	0,02	0,01	0,01	0,01	0,01	0,01	0,01	0,01	0,01	0,01	0,01	0,01	0,02	0,02	
-0,24		0,02	0,02	0,01	0,01	0,01	0,01	0,01	0,01	0,01	0,01	0,01	0,01	0,01	0,02	0,02	
-0,25		0,02	0,02	0,01	0,01	0,01	0,01	0,01	0,01	0,01	0,01	0,01	0,01	0,01	0,02	0,02	
-0,26		0,02	0,02	0,01	0,01	0,01	0,01	0,01	0,01	0,01	0,01	0,01	0,01	0,01	0,02	0,02	
-0,27		0,02	0,02	0,01	0,01	0,01	0,01	0,01	0,01	0,01	0,01	0,01	0,01	0,01	0,02	0,02	
-0,28		0,02	0,02	0,01	0,01	0,01	0,01	0,01	0,01	0,01	0,01	0,01	0,01	0,01	0,02	0,02	
-0,29		0,02	0,02	0,01	0,01	0,01	0,01	0,01	0,01	0,01	0,01	0,01	0,01	0,01	0,02	0,02	
-0,3		0,01	0,02	0,01	0,01	0,01	0,01	0,01	0,01	0,01	0,01	0,01	0,01	0,01	0,02	0,01	
-0,31		0,01	0,02	0,01	0,01	0,01	0,01	0,01	0,01	0,01	0,01	0,01	0,01	0,01	0,02	0,01	
-0,32		0,01	0,02	0,01	0,01	0,01	0,01	0,01	0,01	0,01	0,01	0,01	0,01	0,01	0,02	0,01	
-0,33		0,01	0,02	0,01	0,01	0,01	0,01	0,01	0,01	0,01	0,01	0,01	0,01	0,01	0,02	0,01	
-0,34		0,01	0,01	0,01	0,01	0,01	0,01	0,01	0,01	0,01	0,01	0,01	0,01	0,01			
-0,35		0,01	0,01	0,01	0,01	0,01	0,01	0,01	0,01	0,01	0,01	0,01	0,01	0,01			
-0,36		0,01	0,01	0,01	0,01	0,01	0,01	0,01	0,01	0,01	0,01	0,01	0,01	0,01			
-0,37		0,01	0,01	0,01	0,01	0,01	0,01	0,01	0,01	0,01	0,01	0,01	0,01	0,01			
-0,38		0,01	0,01	0,01	0,01	0,01	0,01	0,01	0,01	0,01	0,01	0,01	0,01	0,01			
-0,39		0,01	0,01	0,01	0,01	0,01	0,01	0,01	0,01	0,01	0,01	0,01	0,01	0,01			
-0,4		0,01	0,01	0,01	0,01	0,01	0,01	0,01	0,01	0,01	0,01	0,01	0,01	0,01			
-0,41		0,01	0,01	0,01	0,01	0,01	0,01	0,01	0,01	0,01	0,01	0,01	0,01	0,01			
-0,42		0,01	0,01	0,01	0,01	0,01	0,01	0,01	0,01	0,01	0,01	0,01	0,01	0,01			

-0,43	0,01	0,01	0,01	0,01	0,01	0,01	0,01	0,01	0,01	0,01	0,01	0,01	0,01					
-0,44	0,01	0,01	0,01	0,01	0,01	0,01	0,01	0,01	0,01	0,01	0,01	0,01	0,01					
-0,45	0,01	0,01	0,01	0,01	0,01	0,01	0,01	0,01	0,01	0,01	0,01	0,01	0,01					
-0,46	0,01	0,01	0,01	0,01	0,01	0,01	0,01	0,01	0,01	0,01	0,01	0,01	0,01					
-0,47	0,01	0,01	0,01	0,01	0,01	0,01	0,01	0,01	0,01	0,01	0,01	0,01	0,01					
-0,48	0,01	0,01	0,01	0,01	0,01	0,01	0,01	0,01	0,01	0,01	0,01	0,01	0,01					
-0,49		0,01	0,01	0,01	0,01	0,01	0,01	0,01	0,01	0,01	0,01	0,01	0,01					
-0,5		0,01	0,01	0,01	0,01	0,01	0,01	0,01	0,01	0,01	0,01	0,01	0,01					
-0,51		0,01	0,01	0,01	0,01	0,01	0,01	0,01	0,01	0,01	0,01	0,01	0,01					
-0,52		0,01	0,01	0,01	0,01	0,01	0,01	0,01	0,01	0,01	0,01	0,01	0,01					
-0,53		0,01	0,01	0,01	0,01	0,01	0,01	0,01	0,01	0,01	0,01	0,01	0,01					
-0,54		0,01	0,01	0,01	0,01	0,01	0,01	0,01	0,01	0,01	0,01	0,01	0,01					
-0,55		0,01	0,01	0,01	0,01	0,01	0,01	0,01	0,01	0,01	0,01	0,01	0,01					
-0,56		0,01	0,01	0,01	0,01	0,01	0,01	0,01	0,01	0,01	0,01	0,01	0,01					
-0,57		0,01	0,01	0,01	0,01	0,01	0,01	0,01	0,01	0,01	0,01	0,01	0,01					
-0,58		0,01	0,01	0,01	0,01	0,01	0,01	0,01	0,01	0,01	0,01	0,01	0,01					
-0,59		0,01	0,01	0,01	0,01	0,01	0,01	0,01	0,01	0,01	0,01	0,01	0,01					
-0,6		0,01	0,01	0,01	0,01	0,01	0,01	0,01	0,01	0,01	0,01	0,01	0,01					
-0,61			0,01	0,01	0,01	0,01	0,01	0,01	0,01	0,01	0,01	0,01	0,01					
-0,62			0,01	0,01	0,01	0,01	0,01	0,01	0,01	0,01	0,01	0,01	0,01					
-0,63			0,01	0,01	0,01	0,01	0,01	0,01	0,01	0,01	0,01	0,01	0,01					
-0,64			0,01	0,01	0,01	0,01	0,01	0,01	0,01	0,01	0,01	0,01	0,01					
-0,65			0,01	0,01	0,01	0,01	0,01	0,01	0,01	0,01	0,01	0,01	0,01					
-0,66			0,01	0,01	0,01	0,01	0,01	0,01	0,01	0,01	0,01	0,01	0,01					
-0,67			0,01	0,01	0,01	0,01	0,01	0,01	0,01	0,01	0,01	0,01	0,01					
-0,68			0,01	0,01	0,01	0,01	0,01	0,01	0,01	0,01	0,01	0,01	0,01					
-0,69			0,01	0,01	0,01	0,01	0,01	0,01	0,01	0,01	0,01	0,01	0,01					
-0,7			0,01	0,01	0,01	0,01	0,01	0,01	0,01	0,01	0,01	0,01	0,01					
-0,71				0,01	0,01	0,01	0,01	0,01	0,01	0,01	0,01	0,01	0,01					
-0,72				0,01	0,01	0,01	0,01	0,01	0,01	0,01	0,01	0,01	0,01					
-0,73				0,01	0,01	0,01	0,01	0,01	0,01	0,01	0,01	0,01	0,01					
-0,74				0,01	0,01	0,01	0,01	0,01	0,01	0,01	0,01	0,01	0,01					
-0,75				0,01	0,01	0,01	0,01	0,01	0,01	0,01	0,01	0,01	0,01					
-0,76				0,01	0,01	0,01	0,01	0,01	0,01	0,01	0,01	0,01	0,01					
-0,77				0,01	0,01	0,01	0,01	0,01	0,01	0,01	0,01	0,01	0,01					
-0,78				0,01	0,01	0,01	0,01	0,01	0,01	0,01	0,01	0,01	0,01					
-0,79						0,01	0,01	0,01	0,01	0,01								
-0,8						0,01	0,01	0,01	0,01	0,01								
-0,81						0,01	0,01	0,01	0,01	0,01								
-0,82						0,01	0,01	0,01	0,01									
-0,83							0,01	0,01	0,01									
-0,84							0,01	0,01	0,01									
-0,85							0,01	0,01	0,01									
-0,86								0,01										
-0,87																		
-0,88																		
-0,89																		
-0,9																		
-0,91																		
-0,92																		
-0,93																		
-0,94																		
-0,95																		
-0,96																		
-0,97																		
-0,98																		
-0,99																		
-1																		
$\sum\limits_{\Delta\vec{h}_x} H_{\beta i/x}$	1,57	1,57	1,56	1,54	1,53	1,52	1,49	1,48	1,48	1,48	1,49	1,52	1,53	1,54	1,56	1,57	1,57	1
$n_{\beta,\Delta\vec{h}_x} =$	35	67	97	121	141	157	165	171	173	171	165	157	141	121	97	67	35	1
$\bar{h}_{\beta/x} =$	0,04	0,02	0,02	0,01	0,01	0,01	0,01	0,01	0,01	0,01	0,01	0,01	0,01	0,01	0,02	0,02	0,04	1
$d_{\beta/x} =$	0,06	0,03	0,02	0,02	0,01	0,01	0,01	0,01	0,01	0,01	0,01	0,01	0,01	0,02	0,02	0,03	0,06	1
$\bar{h}_\beta/d_\beta =$	0,78	0,8	0,8	0,82	0,83	0,84	0,85	0,85	0,86	0,85	0,85	0,84	0,83	0,82	0,8	0,8	0,78	1

$$\bar{\bar{h}}_{k\beta/x(\Delta\vec{h}_x)} = 0{,}07; \quad \vec{\bar{h}}_{k/x,\Delta\vec{h}_x} = 0{,}01; \quad \bar{\bar{h}}_{k\beta/x(\Delta\vec{h}_x)}/\bar{d}_{k/x} = 0{,}834 \rightarrow \pi/4 = 0{,}785398$$

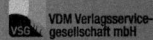